Advancing Maths for AQA
STATISTICS 2

Roger Williamson

Series editors
Roger Williamson   Sam Boardman   Graham Eaton
Ted Graham   Keith Parramore

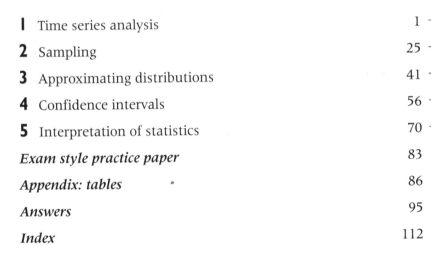

Heinemann Educational Publishers
a division of Heinemann Publishers (Oxford) Ltd,
Halley Court, Jordan Hill, Oxford OX2 8EJ

OXFORD   MELBOURNE   AUCKLAND   JOHANNESBURG
BLANTYRE   GABORONE   PORTSMOUTH NH (USA)   CHICAGO

First published in 2001

01 10 9 8 7 6 5 4 3 2 1

ISBN 0 435 51313 3

Typeset and illustrated by Tech-Set Limited, Gateshead, Tyne & Wear

Printed and bound by Scotprint in the UK

Acknowledgements
The publishers and authors acknowledge the work of the writers, David Cassell,
Ian Hardwick, Mary Rouncefield, David Burghes, Ann Ault and Nigel Price of
the *AEB Mathematics for AS and A-Level Series*, from which some exercises and
examples have been taken.

The publishers' and authors' thanks are due to the AQA for permission to
reproduce questions from past examination papers.

The answers have been provided by the authors and are not the responsibility
of the examining board.

About this book

This book is one in a series of textbooks designed to provide you with exceptional preparation for AQA's new Advanced GCE Specification B. The series authors are all senior members of the examining team and have prepared the textbooks specifically to support you in studying this course.

Finding your way around

The following are there to help you find your way around when you are studying and revising:

- **edge marks** (shown on the front page) – these help you to get to the right chapter quickly;
- **contents list** – this identifies the individual sections dealing with key syllabus concepts so that you can go straight to the areas that you are looking for;
- **index** – a number in bold type indicates where to find the main entry for that topic.

Key points

Key points are not only summarised at the end of each chapter but are also boxed and highlighted within the text like this:

> Time series are analysed so that they may be projected into the future to make forecasts.

Exercises and exam questions

Worked examples and carefully graded questions familiarise you with the specification and bring you up to exam standard. Each book contains:

- Worked examples and Worked exam questions to show you how to tackle typical questions; Examiner's tips will also provide guidance;
- Graded exercises, gradually increasing in difficulty up to exam-level questions, which are marked by an [A];
- Test-yourself sections for each chapter so that you can check your understanding of the key aspects of that chapter and identify any sections that you should review;
- Answers to the questions are included at the end of the book.

CHAPTER 1

Time series analysis

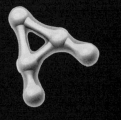

Learning objectives

After studying this chapter, you should be able to:

■ understand the concepts of trend, seasonal variation, short-term variation and random variation
■ describe a trend
■ use moving averages to estimate seasonal effects
■ make forecasts by extrapolating the trend and, where appropriate, applying a seasonal effect
■ modify forecasts, where appropriate, to allow for short-term variation
■ understand that forecasts are merely projections of past patterns and should be treated with caution.

1.1 Introduction

As the name implies a time series is the result of recording a variable at (preferably regular) intervals of time.

For example, daily maximum temperatures, weekly takings of a corner shop or annual profits of large company can be recorded. The following time series is the annual number of marriages in the United Kingdom in thousands.

Year	1987	1988	1989	1990	1991	1992	1993	1994	1995	1996	1997
Marriages in UK	398	394	392	375	350	356	342	331	322	318	310

Unlike most other topics in the syllabus there is no attempt to obtain a random sample of observations. Rather you look for patterns in the data.

> Time series are analysed so that they may be projected into the future to make forecasts.

There is probably more interest in time series analysis than in any other branch of statistics. This is because all organisations need to make forecasts. Governments need to forecast the number of children who will require school places in future years, electricity manufacturers need to forecast the demand for electricity and pub managers need to forecast the demand for beer.

Time series analysis is concerned with examining the past behaviour of a time series to see if patterns can be discerned. These patterns can then be projected into the future. Unfortunately this does not mean that we can tell what is going to happen in the future. It only means that we can say what will happen if current patterns continue.

> Patterns often do not continue and it is foolish to make exaggerated claims for the reliability of any method of forecasting.

There are however many reasons why it may be useful to make a forecast based on past data. These include the following.

- Having a useful starting point for further discussion. For example the pub manager may know that past trends indicate that 220 bottles of Sunderland Brown Ale will be needed to meet next week's demand. However the manager also knows that an advertising campaign is about to start and so modifies the forecast to 280 bottles.
- Having a useful safeguard against overoptimistic (or overpessimistic) forecasts. A sales manager's forecast which is well in excess of current patterns will need to be closely scrutinised.
- A large supermarket chain will need to forecast the sales of thousands of different items. It will need to have a standard method of making these forecasts.
- Comparing actual results (say sales of chocolate bars) with those predicted from past patterns will enable changes in the pattern, and thus the need for a modified method of forecasting, to be identified.
- It is useful for setting targets. A second division football club may aim to attract more spectators to home matches than would be expected from current trends.
- It may draw attention to the fact that some trends are certain to change and the only question is how. For example it is impossible for the world population to continue to increase at its current rate. If it does we will all be standing on each other's toes. The only question is can human society bring about this change in a humane manner or is it to be left to wars, famines and other horrific events?

Time series may usefully be thought of as being made up of some or all of the following four components:

- trend
- seasonal variation
- short-term non-random variation
- random variation.

These components cannot be identified with certainty but nevertheless provide a useful framework for examining time series.

1.2 Trend

A trend is a long-term smooth movement.

The meaning of long-term and short-term is entirely subjective and depends on the circumstances. A meteorologist studying climatic changes may regard long-term as 1000 years and short-term as 20 years; a speculator on the stock exchange may regard long-term as 6 months and short-term as 2 hours.

For example the weekly takings, in £, of a greengrocer over a 12-week period were:

890 900 910 920 930 940

950 960 970 980 990 1000

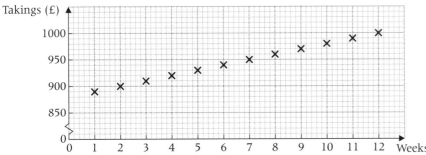

Greengrocer's takings

The vertical scale does not start at zero. This enables the trend to be seen more clearly. However it does exaggerate the trend. If the vertical scale does not start at zero this should always be indicated on the axis.

The series shows an upward linear trend. It is easy to extend this trend to forecast that next week the takings will be £1010. However there is no guarantee that this will be correct.

It is obvious that the series above is artificial. Real data, such as the UK marriage data shown below, does not behave with such regularity. Indeed the word trend implies that short-term or random deviations from this pattern should be ignored.

Worked example 1.1

(a) Plot a graph of the annual number of marriages in the UK from 1987 to 1997 and describe the trend.

Year	1987	1988	1989	1990	1991	1992	1993	1994	1995	1996	1997
Marriages in UK (thousand)	398	394	392	375	350	356	342	331	322	318	310

Source: Office for National Statistics.

(b) The following table shows marriages in the UK classified by the age of the partners.

Numbers

	1987	1988	1989	1990	1991	1992	1993	1994	1995	1996	1997
Males:											
Under 21 years	24 269	20 608	19 070	15 930	13 271	11 031	8 767	7 091	6 302	5 497	5 126
21–24	118 355	109 482	102 977	92 270	79 877	74 458	65 129	56 877	48 432	42 488	36 875
25–29	119 808	120 939	123 491	122 800	115 637	118 255	114 101	111 108	105 218	101 647	97 345
30–34	51 389	53 865	56 442	56 966	56 970	62 470	63 848	65 490	68 245	69 867	70 904
35–44	48 598	51 329	51 411	49 984	48 147	51 125	50 553	51 310	53 350	56 513	58 292
45–54	19 788	21 544	22 329	21 996	20 915	23 290	23 841	24 136	24 786	26 252	26 472
55 and over	15 730	16 282	16 322	15 464	14 922	15 384	15 369	15 220	14 918	15 250	15 204
Females:											
Under 21 years	68 629	59 284	54 256	45 626	38 305	32 618	26 839	22 903	20 643	18 485	17 254
21–24	140 509	134 122	128 411	119 037	105 505	102 494	93 125	84 171	75 071	66 191	59 549
25–29	90 911	95 338	100 531	103 209	99 851	105 223	104 517	102 803	100 644	99 651	97 932
30–34	36 643	39 680	41 989	42 794	43 617	48 514	49 546	52 359	54 819	57 752	58 589
35–44	36 978	39 534	40 290	38 983	37 582	40 075	40 090	41 213	43 115	45 969	47 267
45–54	15 001	16 570	17 172	16 825	16 473	18 504	18 800	19 280	19 720	21 025	21 038
55 and over	9 260	9 521	9 393	8 936	8 406	8 585	8 691	8 503	8 239	8 441	8 589

Source: Office for National Statistics.

Plot the number of marriages and briefly describe the trend for:

(i) Males aged under 21

(ii) Males aged 35–44.

Solution

(a)

Marriages in the UK

There is a downward, approximately linear trend. The figure for 1991 is below the trend.

Pick out the main features. Do not describe every tiny detail of the graph.

(b) (i)

Marriages in the UK – males under 21 years

There is a downward non-linear trend. The rate of reduction is decreasing.

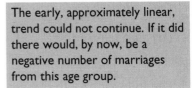

The early, approximately linear, trend could not continue. If it did there would, by now, be a negative number of marriages from this age group.

(ii)

Marriages in the UK – males 35–44

From 1987 to 1994 the number of marriages fluctuates about a horizontal line (that is there is no trend). From 1994 there appears to be an upward trend.

Unlike the younger age groups where the trend is downwards.

EXERCISE 1A

1 The following table shows the number of cinema screens in Great Britain and the total number, in millions, of cinema admissions.

Year	1988	1989	1990	1991	1992	1993	1994	1995	1996	1997	1998
Screens	1117	1177	1331	1544	1547	1591	1619	1620	1738	1886	1975
Admissions	75.2	82.9	78.6	88.9	89.4	99.3	105.9	96.9	118.7	128.2	123.4

Source: Office for National Statistics.

(a) (i) Plot the data for number of screens and briefly describe the trend.

(ii) Plot the data for admissions and briefly describe the trend.

(b) Compare the two trends.

2 The following table shows the number of notifications, in hundreds, of measles and of food poisoning in the United Kingdom.

Year	1988	1989	1990	1991	1992	1993	1994	1995	1996	1997	1998
Measles	906	310	156	117	123	120	235	90	69	48	45
Food poisoning	457	592	597	595	721	767	911	926	949	1056	1050

Source: Annual Abstract of Statistics, 2000.

Plot the data and briefly describe the trend for:

(a) measles

(b) food poisoning.

3 The following table shows the population, in thousands, of the City of Manchester and of Greater Manchester.

Year	1911	1931	1951	1961	1971	1981	1991	1998
Manchester City	714	766	703	657	554	463	439	430
Greater Manchester	2638	2727	2716	2710	2750	2619	2570	2577

Source: Census and Office for National Statistics.

Plot the data and briefly describe the trend for:

(a) the City of Manchester

(b) Greater Manchester.

Compare the two trends.

1.3 Seasonal variation

This is the most readily understood component of a time series. We all expect more electricity to be used in winter than in summer and more ice creams to be sold in summer than in winter. Seasonal effects do not necessarily refer to the seasons of the year. For example a greengrocer's takings are likely to be higher on Friday and Saturday than on Monday. This is also called a seasonal effect.

A seasonal effect is a regular predictable pattern.

A market stall is open on Tuesday, Friday and Saturday each week. The takings, in £, over a 4-week period are:

T	F	S	T	F	S
320	525	580	335	540	595

T	F	S	T	F	S
350	555	610	365	570	625

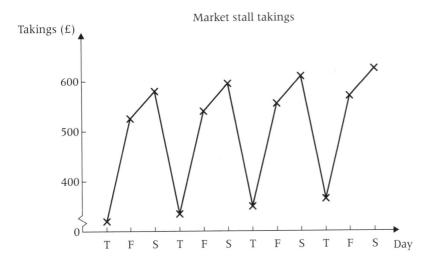

Market stall takings

> Joining the points with straight lines makes the seasonal pattern clearer.

> Don't forget to indicate that the vertical scale does not start at zero.

The graph shows that on Tuesdays takings are low and on Fridays and Saturdays they are high. In other words there is a seasonal effect. To analyse the series it is useful to attempt to remove the seasonal effect (called deseasonalising) and analyse what remains of the series. If a forecast is to be made it will be made from the deseasonalised data and the seasonal effect will then be added back in.

There are many different ways of deseasonalising data and as with many things in statistics it is not possible to say one is right and another is wrong. However the most straightforward way is to use moving averages. In the example above the seasonal effect occurs over the 3 days in a week on which the market stall is open. That is the seasonal pattern repeats itself every three observations. For this reason the observations are compared with a three-point average. The mean of the first three observations is $\frac{(320 + 525 + 580)}{3} = 475$. This should be compared with the middle of the three observations used, that is, with 525. This observation is 50 above the moving average. The next moving average is $\frac{(525 + 580 + 335)}{3} = 480$. This should be compared with 580. This observation is 100 above the moving average.

> If the first moving average was compared with, say, the observation for the first Tuesday, it would not be possible to tell whether a difference was due to a seasonal effect or due to a trend.

Day	T	F	S	T	F	S	T	F	S	T	F	S
Takings	320	525	580	335	540	595	350	555	610	365	570	625
Moving average		475	480	485	490	495	500	505	510	515	520	

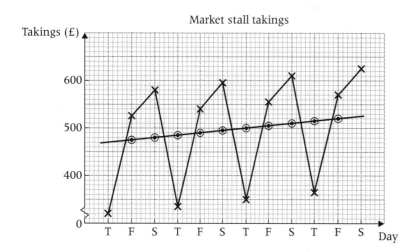

Market stall takings

> The seasonal effect may be removed by calculating a suitable moving average.

The moving average shows an upward linear trend. It is easy to extend this trend and predict that for the next week the moving average will be:

T	F	S
530	535	540

To predict the actual takings for the next week a numerical value of the seasonal effect is required. The following table shows the actual observations minus the moving average.

Day	T	F	S	T	F	S	T	F	S	T	F	S
Takings	320	525	580	335	540	595	350	555	610	365	570	625
Moving average		475	480	485	490	495	500	505	510	515	520	
Takings – moving average		50	100	−150	50	100	−150	50	100	−150	50	

On Fridays the takings are always 50 above the moving average, on Saturdays they are 100 above the moving average and on Tuesdays they are 150 below the moving average.

The predictions for next week are therefore:

Tuesday	$530 - 150 = 380$
Friday	$535 + 50 = 585$
Saturday	$540 + 100 = 640$.

Notice that it is not possible to calculate a moving average corresponding to the first Tuesday or the last Saturday.

The moving averages are shown as ⊙

The sign is important.

It is easy to make a mistake and subtract the seasonal effect when you should have added it. If you do make this mistake it will be obvious your predictions do not follow the earlier pattern of low on Tuesday, high on Friday and Saturday.

Real data

The time series above was clearly fictitious. Real data would not behave with such regularity. Here are some real data. The table shows the daily number of requests, in thousands, for pages from a government department's website over a 3-week period.

Sunday	84	Thursday	151
Monday	162	Friday	178
Tuesday	192	Saturday	74
Wednesday	189	Sunday	84
Thursday	171	Monday	176
Friday	169	Tuesday	149
Saturday	76	Wednesday	181
Sunday	86	Thursday	192
Monday	156	Friday	171
Tuesday	194	Saturday	78
Wednesday	190		

There is clearly a pattern to the weekly requests with the most notable feature being the reduction in requests on Saturdays and Sundays. Since the pattern recurs every 7 days the observations are compared with a seven-point moving average.

The first moving average will be compared with the fourth point.

Day	Requests (thousands)	Moving average (MA)	Requests – MA
Sunday	84		
Monday	162		
Tuesday	192		
Wednesday	189	149.0	40.0
Thursday	171	149.3	21.7
Friday	169	148.4	20.6
Saturday	76	148.7	−72.7
Sunday	86	148.9	−62.9
Monday	156	146.0	10.0
Tuesday	194	147.3	46.7
Wednesday	190	147.0	43.0
Thursday	151	146.7	4.3
Friday	178	149.6	28.4
Saturday	74	143.1	−69.1
Sunday	84	141.9	−57.9
Monday	176	147.7	28.3
Tuesday	149	146.7	2.3
Wednesday	181	147.3	33.7
Thursday	192		
Friday	171		
Saturday	78		

It is not possible to calculate a seven-point moving average corresponding to the first three or the last three points.

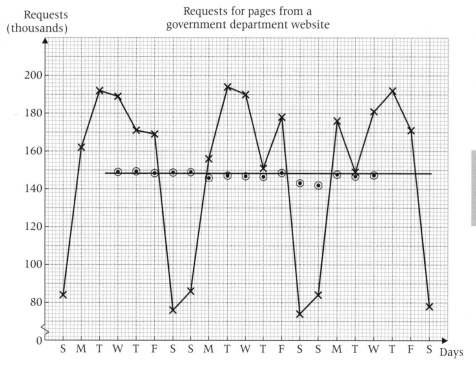

Requests
(thousands)

Requests for pages from a
government department website

S M T W T F S S M T W T F S S M T W T F S Days

With this real data there is still a seasonal pattern but it is less regular than the artificial data.

From the graph you can see that the moving average is very close to horizontal – in other words there is no trend.

However if you look at the difference between the actual number of requests on, say, Wednesdays and the moving average you see that it is 40.0, 43.0 and 33.7. These three numbers are not the same but they are similar. The best estimate that can be made for the seasonal effect for Wednesdays is the mean of these three numbers.

That is $\dfrac{(40.0 + 43.0 + 33.7)}{3} = 38.9$

For the other days of the week only two moving averages are available and so the best that can be done is to average the (signed) deviation from the moving average for these 2 days.

	Mon	Tues	Wed	Thur	Fri	Sat	Sun
			40.0	21.7	20.6	−72.7	−62.9
	10.0	46.7	43.0	4.3	28.4	−69.1	−57.9
	28.3	2.3	33.7				
Estimated seasonal effect	19.1	24.5	38.9	13.0	24.5	−70.9	−60.4

The sign is important.

This is the average of the numbers in the column above. If there was more data we could improve the estimate.

The graph indicates that there is no noticeable trend in the moving average and so to forecast the number of requests for the next week you could start with the current value of the moving average of 147.3 and apply the estimated seasonal effects. Monday has been estimated to be 19.1 above the moving average and so our forecast for Monday would be:

$$147.3 + 19.1 = 166$$

It is reasonable to give three significant figures in this case. If nothing happens to disturb the current pattern the past data suggests that you cannot hope for the forecasts to be accurate to 3 sf. You could hope for most of them to be accurate to 2 sf. To give more than 3 sf suggests that you are claiming unattainable levels of accuracy for your forecasts.

To complete the forecasts for the week:

Tuesday	$147.3 + 24.5 = 172$
Wednesday	$147.3 + 38.9 = 186$
Thursday	$147.3 + 13.0 = 160$
Friday	$147.3 + 24.5 = 172$
Saturday	$147.3 - 70.9 = 76$
Sunday	$147.3 - 60.4 = 87$

Worked example 1.2

A market stall holder has been selling petfood on Thursday, Friday and Saturday each week for several years. The stall holder employs a consultant to analyse the business. The consultant collects the following data of takings for the last 3 weeks.

Day	1	2	3	4	5	6	7	8	9
	Thur	Fri	Sat	Thur	Fri	Sat	Thur	Fri	Sat
Takings (£)	278	396	592	312	409	622	315	431	621

(a) Plot the data together with a suitable moving average.　[A]

(b) Use regression analysis to predict the value of the moving average for the Saturday of the next week.　[A]

(c) Based on your answer in part **(b)** predict the takings for the Saturday of the next week.　[A]

(d) The consultant reports that the stall will be taking about £1450 on Saturday in a years' time and this will rise to £2330 on a Saturday in 2 years' time. Comment on the consultant's report.

Solution

(a)

Day	1	2	3	4	5	6	7	8	9
	278	396	592	312	409	622	315	431	621
MA		422	433.3	437.7	447.7	448.7	456	455.7	

> A three-point moving average is appropriate.

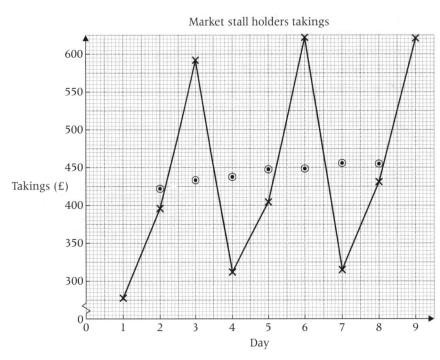

Market stall holders takings

(b) Since the question specifies regression should be used you need to regress the moving average (y) on the number of the day (x). Since the moving averages only run from day 2 to day 8 only these 7 days can be used.

The data can be entered into a calculator to give:

$$y = 415 + 5.62x$$

For the next week Thursday will be day 10, Friday will be day 11 and Saturday will be day 12. The predicted moving average for Saturday of the next week will be:

$$415 + 5.62 \times 12 = 482$$

> Note. In this question the days were numbered for you. If they had not been you would have had to number them yourself. The given numbering is probably the simplest but there is no reason why you cannot start with zero or any other number.

(c) The differences between the takings and the moving averages on the two Saturdays where this can be calculated is:

$$592 - 433.3 = 158.7$$
$$622 - 448.7 = 173.3$$
$$\text{mean} = 166.0.$$

Prediction for Saturday of next week is $482 + 166 = 648$

> There is no moving average available to compare with the third Saturday.

> Alternatively you could calculate the residuals from the regression line.

(d) The consultant has apparently extrapolated 3 weeks data to 1 year and 2 years ahead. It is foolish to extrapolate so little data so far ahead. The number of significant figures gives a wholly spurious impression of the likely accuracy of the predictions.

> Do not give more than 3 sf.

Worked example 1.3

The following table shows the expenditure, in £ million, by UK households on air travel. The figures have been adjusted to constant 1995 prices. These should be used throughout the question.

Year	1997	1998	1999
Quarter 1	1206	1340	1454
Quarter 2	1676	1648	1954
Quarter 3	2139	2123	2530
Quarter 4	1384	1589	1780

Source: Office for National Statistics.

> Examine the table carefully. To arrange the observations in order of time you need to go down the first column, then down the second, etc. In other questions you might have to go across rows.

(a) Plot the data together with a suitable moving average.

(b) Describe the trend shown by the moving average.

(c) By extrapolating the trend shown by the last six moving averages predict the moving average for Quarter 1, 2000.

(d) Predict the actual expenditure on air travel, for Quarter 1, 2000.

(e) The actual expenditure in Quarter 1 was £1646 million. Compare this with your forecast and comment.

(f) Comment briefly on the fact that the data in this question is in constant 1995 prices.

Solution

(a) A four-point moving average is appropriate here. This means that to avoid contaminating the seasonal effect with a possible trend you will need to plot the first moving average halfway between Quarter 1 and Quarter 2.

> This will always occur if there are an even number of points in the moving average.

> The moving average has been shown on alternate lines to indicate that it corresponds to a point halfway between the two adjoining quarters.

		Expenditure £million	Moving average
1997	Q1	1206	
	Q2	1676	
			1601
	Q3	2139	
			1635
	Q4	1384	
			1628
1998	Q1	1340	
			1624
	Q2	1648	
			1675
	Q3	2123	
			1703
	Q4	1589	
			1780
1999	Q1	1454	
			1882
	Q2	1954	
			1929
	Q3	2530	
	Q4	1780	

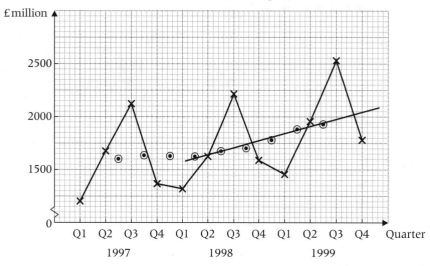

Expenditure by UK households on
air travel, constant 1995 prices

(b) The moving average starts off horizontally (i.e. there is no trend) but then shows a small, approximately linear, upward trend.

(c) Moving average for Quarter 1, 2000 about £2100 million.

(d) You need to compare the Quarter 1 figures for 1998 and 1999 with the moving average. This cannot be done directly from the table. It is perfectly acceptable to do this graphically. It can also be done numerically by taking the mean of the two adjoining moving averages.

The moving average immediately before Quarter 1, 1998 is 1628 and the one immediately after is 1624.

The mean is $\frac{(1628 + 1624)}{2} = 1626$. This is sometimes called the centred moving average.

The observed value for Quarter 1, 1998 is 1340, which is $1340 - 1626 = -286$ from the moving average.

Making the same calculation for Quarter 1, 1999 gives a centred moving average of $(1780 + 1882)/2 = 1831$. The observed value of 1454 is $1454 - 1831 = -377$ from the moving average.

Taking the mean of -286 and -377 gives -331.5.

The prediction for Quarter 1, 2000 is $2100 - 331.5 = 1770$.

If there are more data or if the seasonal effects for the other quarters are required it is worth putting the calculations in a table.

		Expenditure £million	Moving average	Centred moving average (CMA)	Moving average −CMA
1997	Q1	1206			
	Q2	1676			
			1601		
	Q3	2139			
			1635		
	Q4	1384			
			1628		
1998	Q1	1340		1626	−286
			1624		
	Q2	1648			
			1675		
	Q3	2123			
			1703		
	Q4	1589			
			1780		
1999	Q1	1454		1831	−377
			1882		
	Q2	1954			
			1929		
	Q3	2530			
	Q4	1780			

These are the means of the adjoining moving averages. If you were asked to estimate the seasonal effect for all quarters you would need to complete the last two columns.

(e) The actual value is substantially less than that predicted. Not too much should be read into the figures for one quarter but this suggests that the upward trend of the immediately preceding quarters may have slowed down.

(f) The fact that the data is given at constant 1995 prices means that an upward trend represents a real increase in expenditure on air travel. If the data had not been at constant prices the upward trend might merely reflect inflation.

EXERCISE 1B

1 A mobile fish shop calls at a village on Mondays, Wednesdays and Fridays. The value of the fish sold over a 3-week period was as follows:

Day	M	W	F	M	W	F	M	W	F
Sales, £	17	31	59	20	27	63	21	23	66

Plot the data together with an appropriate moving average.

2 A theatre opens from Tuesday to Saturday. The theatre holds a maximum of 600 people. The attendances for the first 3 weeks of a new play are shown in the table below.

	Week 1					Week 2					Week 3			
T	W	T	F	S	T	W	T	F	S	T	W	T	F	S
199	216	230	320	430	228	209	250	347	470	254	276	296	402	504

(a) Plot a graph of the data together with a suitable moving average.

(b) A severe storm occurred shortly before one of the performances. Which performance do you think this was? Give a reason.

(c) Predict the attendance on the Tuesday of week 4.

(d) Explain why the trend shown by the moving average cannot continue over a long period of time.

3 The quarterly sales (£ thousands) of an agricultural chemical company over a 2-year period are shown below.

	1999				2000		
Q1	Q2	Q3	Q4	Q1	Q2	Q3	Q4
165	170	125	149	x	$1.2x$	130	149

An appropriate moving average is calculated.

(a) Find the value of the first point.

(b) Given that the value of the second point is 160 find the value of the third point.

4 The following table shows the expenditure, in £ million, of UK households on horticultural goods. The figures have been adjusted to constant 1995 prices.

Year	1997	1998	1999
Quarter 1	579	629	739
Quarter 2	857	925	1056
Quarter 3	493	482	632
Quarter 4	411	489	550

Source: Annual Abstract of Statistics, 2000.

(a) Plot the data together with a suitable moving average.

(b) Describe, briefly, the trend shown by the moving average.

(c) Predict the expenditure for Quarter 1, 2000 based on the data above.

(d) The actual expenditure for Quarter 1, 2000 was £791 million. Compare this with the forecast you made in part **(c)** and comment.

(e) Suggest a reason for the seasonal effects you have observed.

5 The expenditure, in £ million, of UK households on sports and toys is shown in the following table. The figures have been adjusted to constant 1995 prices.

Year	Quarter 1	Quarter 2	Quarter 3	Quarter 4
1997	1273	1474	1492	2094
1998	1552	1870	1869	2507
1999	1789	2195	2244	3039

Source: Annual Abstract of Statistics, 2000.

(a) Plot the data together with a suitable moving average.

(b) Describe the trend shown by the moving average.

(c) Use regression analysis to predict the moving average for each quarter of 2000.

(d) Predict the actual expenditure for each quarter of 2000.

(e) The actual expenditure in the first quarter of 2000 was £2080 million. Compare this with your forecast and comment.

(f) Suggest a possible reason for the seasonal effects you have observed in the data.

1.4 Random variation

It is clear from the examples above which involve real data that time series do not behave in an entirely regular or predictable manner. This unpredictable component of the series is called random variation. The series illustrated below exhibits random variation about an upward linear trend.

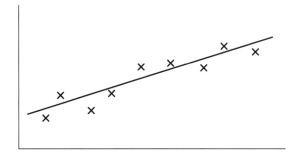

It is impossible to forecast random variation. If it can be forecast then it is not random.

 All real time series will contain some random variation which is impossible to forecast.

1.5 Short-term variation

The time series below has a long-term upward linear trend but varies around this trend in a manner which is neither regular nor random.

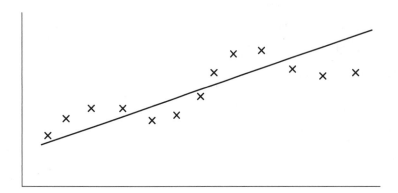

This non-random variation about a trend is called short-term variation.

 Short-term variation is variation about the trend which is neither regular nor random.

Worked example 1.4

Plot points illustrating a time series which displays:

(a) random variation about a downward linear trend

(b) short-term variation about a downward linear trend

(c) seasonal variation but no trend

(d) random variation about a downward non-linear trend.

Solution

(a)

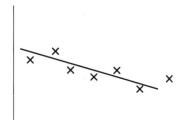

(b)

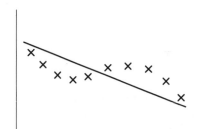

(c)

(d)

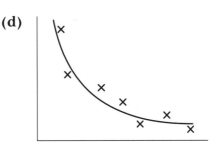

Worked example 1.5

The following data is the annual membership of a rugby club.

Year	1987	1988	1989	1990	1991	1992	1993	1994	1995	1996	1997	1998	1999	2000
Number of members	132	142	149	148	146	144	162	179	196	204	221	233	253	261

(a) Plot the data.

(b) Calculate the equation of the regression line of number of members, y, on year, x, and draw the line on your graph.

(c) Describe the behaviour of the series.

(d) Use the regression line to estimate the number of members of the club in 2001.

(e) Examine your graph and modify the estimate you have made in part **(d)**.

Solution

(a)

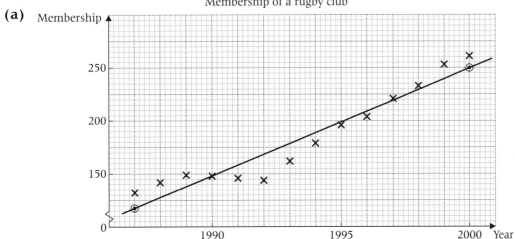

Membership of a rugby club

(b) $y = -20\,093.2 + 10.1714x$ from calculator

> The raw data has been entered uncoded. If you do this a large number of significant figures are needed because any rounding error in the gradient will be multiplied by about 2000.
>
> Alternatively you can code the data, 1987 → 1, 1988 → 2, etc. before undertaking the calculation.

$x = 1987, \hat{y} = 117.5 \quad x = 2000, \hat{y} = 249.7$

(c) The series shows short-term variability about an upward linear trend.

(d) $x = 2001$ $\hat{y} = 260$.

Regression line estimates a membership of 260 for 2001.

(e) The graph suggests that the actual number will be above the regression line. Modified estimate, say, 270.

EXERCISE 1C

1 Plot points illustrating a time series which displays:

(a) random variation about an upward linear trend

(b) seasonal variation

(c) short-term variation about a downward linear trend

(d) random variation about a downward non-linear trend

(e) random variation but no trend

(f) short-term variation about an upward non-linear trend.

2 **A, B, C, D, E** and **F** are descriptions of time series.

A random variation about a linear trend

B seasonal variation about a linear trend

C random variation about a downward non-linear trend

D short-term variation about an upward linear trend

E short-term variation about a linear trend

F short-term variation about a non-linear trend.

Choose the description **A, B, C, D, E** or **F** which best describes each of the time series below.

(a)

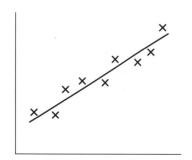

(b)

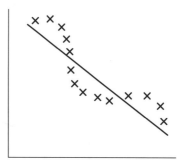

(c)

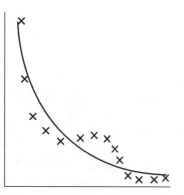

(d)

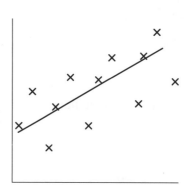

3 The following data is the membership of a lodge of a secretive sexist society. Year 1 is 1986, year 2 is 1987, etc.

Year	1	2	3	4	5	6	7	8	9	10	11	12	13	14	15
Membership	254	239	225	210	206	192	188	172	151	142	120	94	79	59	45

(a) Plot the data.

(b) Calculate the equation of the regression line of membership on year.

(c) Describe the behaviour of the series.

(d) Use the regression line to predict the membership in 2001.

(e) Use your graph to modify your prediction in part (d).

(f) What problems would arise in attempting a prediction for 2005?

MIXED EXERCISE

1 The value of goods exported from the United Kingdom to Norway, £ million, is shown in the following table.

Year	1989	1990	1991	1992	1993	1994	1995	1996	1997	1998
Exports	1057	1291	1323	1421	1502	2035	2007	2066	2659	2757

Source: Annual Abstract of Statistics, 2000.

(a) Plot the data.

(b) Describe the trend.

(c) How would the trend be affected if the figures had been adjusted to allow for inflation?

2 The following table shows the number of consultants and the number of nurses and midwives employed in the National Health Service.

Year	1988	1989	1990	1991	1992	1993	1994	1995	1996	1997	1998
Consultants (hundreds)	489	503	518	530	540	556	567	602	622	643	674
Nurses and midwives (thousands)	490	491	487	484	468	446	429	354	356	354	410

Source: Annual Abstract of Statistics, 2000.

Plot the data and describe the trend for:

(a) consultants

(b) nurses and midwives.

Compare the magnitudes and the trends of the two series.

3 Cecilia takes over a milkround from Alex. She collects money on Tuesdays, Fridays and Saturdays. Her takings for the first 3 weeks were:

Day	T	F	S	T	F	S	T	F	S
Takings (£)	196	340	413	212	373	468	234	400	480

(a) Plot the data together with a suitable moving average.

(b) By extrapolating the trend shown by the moving average predict the moving average for Tuesday, Friday and Saturday of the next week.

(c) Predict the actual takings on Tuesday, Friday and Saturday of the next week.

(d) What reservations would you have about using the method of parts **(b)** and **(c)** to predict the takings for the week, 1 year after she took over the round.

4 The following table gives the oil usage (in tonnes) of a small engineering firm.

Year \ Quarter	1	2	3	4
1997	125	96	72	119
1998	137	113	88	131
1999	117	118	94	142
2000	162	155	162	176

(a) Plot the data together with a suitable moving average.

(b) During one of the quarters the works was shut down for 3 weeks due to a major breakdown. Which quarter do you think this was? Explain your answer.

(c) During another quarter a leak was found in the oil tank which was then replaced. Which quarter do you think this was. Explain your answer.

5 The following table shows the expenditure, £ million, on stationery of households in the UK. The data has been adjusted to constant 1995 prices.

Year	1997	1998	1999
Quarter 1	810	838	854
Quarter 2	780	764	779
Quarter 3	820	812	814
Quarter 4	1063	1065	1029

Source: Consumer Trends, 2000.

(a) Plot the data together with a suitable moving average.

(b) Use regression analysis to predict the moving average for the four quarters of 2000.

(c) Predict the actual expenditure for the four quarters of 2000.

(d) Given that the actual expenditure for the first quarter of 2000 was £859 million, comment on your method of forecasting.

6 The following data is the weekly output of a weaving mill in terms of lengths of cloth produced. The figures have been adjusted for seasonal effects, including bank holidays.

Week	1	2	3	4	5	6	7	8	9	10	11	12	13	14
Lengths	264	283	298	296	292	287	324	338	393	407	442	466	502	532

(a) Plot the data.

(b) Calculate the equation of the regression line of lengths on week. Draw the line on your graph.

(c) Describe the behaviour of the series.

(d) Use the regression line to estimate the number of lengths produced in week 15.

(e) Examine your graph and modify the estimate you have made in part **(d)**.

Key point summary

1 Time series are analysed so that they may be projected into the future to make forecasts.   *p1*

2 There are many good reasons for doing this but past patterns do not always continue and so it is unwise to expect the forecasts to be accurate.   *p2*

3 A trend is a long-term smooth movement. It may be estimated by eye from a graph or, if it is approximately linear, by using regression.   *p3*

4 A seasonal effect is a regular predictable pattern, such as cinema audiences being higher on Fridays and Saturdays than on Mondays.   *p6*

5 The seasonal effect may be removed by calculating a suitable moving average. An estimate of the magnitude of the seasonal effect may be made by comparing each point of a time series with the corresponding moving average.   *p8*

6 All real time series will contain some random variation which is impossible to forecast.   *p17*

7 Short-term variation is variation about the trend which is neither regular nor random.   *p18*

Test yourself	What to review
1 Time series may be usefully be considered as consisting of some or all of four components. What are these components?	*Section 1.1*
2 What is meant by 'long term' in the context of a time series?	*Section 1.2*
3 You are asked to forecast a series which contains only random variation. What problem would arise?	*Section 1.4*
4 A fishmonger opens 5 days a week from Tuesday to Saturday. The table shows the daily takings, to the nearest £10, over a 2-week period.	*Section 1.3*

Day	Tu	W	Th	F	S	Tu	W	Th	F	S
Takings (£)	860	540	690	1020	980	900	580	730	1060	1020

Choose a suitable moving average and calculate the values for the two Thursdays.

5 If the trend of the moving average in question **4** is extended it suggests that the moving average for Thursday of the next week would be £898. Forecast the actual takings for Thursday of the next week.	*Section 1.3*
6 What feature of the data in question **4** suggests that it is fictitious?	*Section 1.3*
7 Why is it not possible to calculate a moving average for the second Friday for the data in question **4**?	*Section 1.3*
8 Sketch a time series which shows short-term variation about an upward linear trend.	*Section 1.5*

Test yourself **ANSWERS**

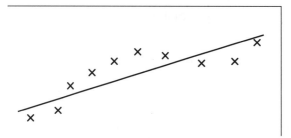

8 This is one possible answer.

7 For a five-point moving average the data for 2 days ahead is necessary. This is not available for the second Friday.

6 The seasonal pattern is exactly the same in both weeks.

5 £770.

4 £818, £858.

3 It is not possible to forecast random variation.

2 Long term is subjective and depends entirely on the context.

1 Trend, seasonal variation, short-term variation, random variation.

CHAPTER 2

Sampling

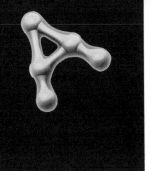

Learning objectives

After studying this chapter, you should be able to:

■ define random, stratified, systematic, quota and cluster samples
■ select an appropriate method of sampling for use in particular circumstances
■ understand the advantages and disadvantages of the different methods of sampling.

2.1 Sampling

The purpose of sampling is to obtain information about a population by examining only part of it – a sample.

This may be done for reasons of economy – it is easier and cheaper to test a sample of components than to test the whole batch. It may be done because it is impossible, in practice, to examine the whole population – for example you would never be able to interview every adult in the United Kingdom. Or it may be done because the test is destructive – it would be ridiculous to find the average working life of a large batch of electric light bulbs by testing them all before selling them.

If a sample is to be used to make inferences about a population or to estimate population parameters it is essential that the sample is chosen, as far as possible, to be representative of the population and to avoid bias.

2.2 Random sampling

Random sampling is defined in S1. To take a random sample of size n from a population of size N, every member of the population must have an equal chance of inclusion. However this alone is not sufficient, it is also necessary that every subset of size n of the population must have an equal chance of forming the sample. Simple random samples are taken without replacement, unrestricted random samples are taken with

> Unrestricted random samples are rarely used as this allows the possibility of the same member of the population being included more than once in the sample.

replacement. Random numbers from tables, calculators or computer packages are usually used in order to select a random sample.

> A random sample of size n is a sample selected in such a way that all possible samples of size n have an equal chance of being selected.

2.3 Stratified sampling

When sampling for opinion polls or social surveys there is usually a substantial amount of information available about the population which is being sampled. This information may come from the census, government publications or local authority records. It may be used to ensure that the sample is representative of the population. For example if we know that 34% of the population is aged over 60 we can ensure that 34% of the sample is aged over 60. This is only useful if the over-60s as a group have different opinions from the rest of the population. If there is no difference from the rest of the population then the extra work needed to take a stratified sample is done to no purpose.

There may be circumstances where it would be desirable to 'over-represent' strata. For example to ensure that more than 34% of the sample is aged over 60. This is beyond the scope of this book.

It is generally understood that stratified sampling implies that a random sample is taken from each of the strata of the population. To emphasise this point this type of sampling is sometimes referred to as stratified random sampling.

This is **not** a random sample since all subsets of the population cannot be chosen. In a random sample it would be possible (but unlikely) for the whole sample to come from the same strata.

> A stratified sample requires prior knowledge to be used to divide the population into strata. Random samples are then taken from each of these strata, usually in proportion to the size of each of the strata.

> If relevant prior information about the population is available, a stratified sample using this information is preferable to a random sample.

Worked example 2.1

A gardener grew 28 tomato plants of which 21 were of strain A and seven of strain B. The gardener wishes to know the weight of tomatoes produced by the plants and decides to estimate this by weighing the yield of eight plants. Describe how random numbers could be used to select a stratified sample of size 8 from the 28 plants.

Solution

Of 28 plants, 21 are of strain A. So the sample should contain a proportion of $\frac{21}{28} = 0.75$ plants of strain A.

That is $0.75 \times 8 = 6$ plants of strain A.

Number the 21 plants of strain A from 00 to 20.

Select two-digit random numbers.

Ignore repeats and numbers greater than 20.

Continue until six two-digit numbers have been obtained. Select the corresponding plants.

To complete the sample we require two plants of strain B.

Number the seven plants of strain B from 0 to 6.

Select random digits, ignoring repeats and numbers greater than six. When two have been obtained select the corresponding plants.

EXERCISE 2A

1 Explain briefly what you understand by a **stratified sample**. Give one advantage and one disadvantage of using stratified sampling compared to simple random sampling.

2 A German class consists of 12 male students and 18 female students. Describe how you would use random numbers to select a stratified sample of size 10 from this class.

3 A survey is to be carried out into the attitudes of voters in a parliamentary constituency to their member of parliament.

   (a) Discuss the advantages and disadvantages of using a stratified sample if there is known to be:
      (i) a difference between the attitudes of men and women
      (ii) no difference between the attitudes of men and women.

   (b) Suggest three factors other than gender which might be used to stratify the population.

2.4 Quota sampling

If a survey of a large population, such as all adults living in the United Kingdom, is undertaken it is usually impossible to achieve a random sample. There are difficulties in identifying the population, locating the individuals chosen and having located them persuading them to answer the questions. A **quota** sample is a stratified sample where no attempt is made to select the sample at random. The interviewers are given a quota of people to locate from given strata but are left to choose for themselves the particular individuals to include in the sample. For example an interviewer might be asked to interview eight male manual workers living in a suburban area.

Quota sampling is not as good as stratified random sampling because the interviewer will inevitably interview people who are easily accessible and willing to be interviewed. These people as a group may have different views from the population as a whole. However, quota sampling has been found to give useful results when the stratification is skillfully carried out.

> A quota sample is a stratified sample where the samples from each of the strata are chosen for convenience and there is no attempt at random sampling.

Worked example 2.2

In a large city 350,000 of the residents draw a state pension. These include 210,000 females of whom 80,000 are aged over 80. Of the males 20,000 are aged over 80.

A sample of 350 residents who draw a state pension is to be asked their views on the services for the elderly provided by the city.

(a) Give two advantages and one disadvantage of using a quota sample compared to a random sample.

(b) How many residents from each of the strata should be included in the sample and how should they be chosen?

Solution

(a) A quota sample is easier to carry out than a random sample (which would probably be impossible in practice).

Each of the strata would be fairly represented in the sample. This is desirable if different strata have different views.

Interviewers will choose people easily accessible to themselves which may introduce bias. A random sample would avoid this problem.

(b) The sample consists of 0.1% of the population and so a quota sample would consist of 0.1% of each of the strata.

i.e 80 females over 80

130 females not over 80

20 males over 80

120 males not over 80.

The samples from each of the strata should be chosen in any way convenient to the interviewer.

> If the numbers were not so convenient it might be impossible to take exactly the same proportion of each of the strata in the sample. It is possible to allow for different proportions from each of the strata in analysing the data. This is beyond the scope of this book.

2

EXERCISE 2B

1 Explain what you understand by a quota sample. Under what circumstances would you recommend the use of quota sampling?

2 A school has 1100 pupils of whom 320 are in the sixth form. A total of 180 of the sixth formers are girls as are 380 of the pupils who are not in the sixth form. A quota sample of size 55 is to be selected and asked to complete a questionnaire. Give detailed instructions to the school secretary who has no knowledge of sampling methods but has to decide which pupils are to be asked to complete the questionnaire.

3 A manufacturing company operates 15 factories, each with different numbers of employees. It has a total of 13 800 employees. The board of directors, concerned by a large turnover, decides to survey 100 employees to seek their opinions on working conditions. The following suggestions were made as to how the sample could be chosen.

Suggestion A. The employees are numbered 00 000 to 13 799. One hundred different five-digit random numbers between 00 000 and 13 799 are taken from random number tables and the corresponding employees chosen.

Suggestion B. The sample is made up of employees from all factories. The employees are selected at random from each factory, the number from each factory being proportional to the number of employees at the factory.

Suggestion C. The sample is made up of employees from all factories. The employees are selected by a convenient method, the number from each factory being proportional to the number of employees at the factory.

(a) (i) Which of the suggestions would produce a **quota** sample? [A]

(ii) Name the type of sampling described in each of the other **two** suggestions. [A]

(b) For each of the **three** suggestions state whether or not all employees have an equal chance of being included in the sample.

(c) Give **one** reason for using:

(i) **Suggestion A** in preference to **Suggestion C**,

(ii) **Suggestion C** in preference to **Suggestion A**.

(d) Explain why **Suggestion B** might be preferred to **Suggestion A**.

2.5 Cluster sampling

Cluster sampling is used where it would be impractical to take a random sample but it is desirable to keep a random element in the sampling method. For example, although it would be theoretically possible to interview a random sample of all junior school teachers in the United Kingdom it would involve an excessive amount of travelling and expense. Instead you could select a random sample of junior schools in the United Kingdom and then interview either all the teachers in these schools or a random sample of the teachers in the selected schools.

The amount of travelling would be greatly reduced. The disadvantage is that the teachers in a particular school are unlikely to hold views which are as varied as those of the population of all junior school teachers.

There can be more than one step to cluster sampling. For example you could start by selecting a random sample of counties and then a random sample of junior schools within the selected counties. This would further cut down the travelling as the selected schools would now be clustered together rather than spread throughout the country.

> Teachers in a particular school are likely to have views which are more **homogeneous** than the population of all teachers. The views of all teachers will be more **heterogeneous**.

> This is called **multi-stage** cluster sampling.

A cluster sample uses randomly chosen clusters of the population. It is not a random sample but contains an element of randomness.

Worked example 2.3

A health service union wishes to distribute a questionnaire to a sample of nurses working in hospitals. A list of all the hospitals in the United Kingdom is obtained and three are selected at random. Twenty nurses are chosen at random from each of the selected hospitals and asked to complete the questionnaire.

(a) What is the name given to this type of sampling? [A]

(b) Give **one** advantage and **one** disadvantage of this method of sampling. [A]

(c) Are all nurses working in hospitals in the United Kingdom equally likely to be included in the sample? Explain your answer. [A]

Solution

(a) Cluster sampling

(b) Advantage – easier to administer as chosen nurses will be clustered in three hospitals instead of spread throughout the country.

Disadvantage – nurses in the same hospital likely to have less varied views than the whole population of nurses.

(c) No, nurses in small hospitals are more likely to be chosen than nurses in large hospitals.

> The **hospitals** have an equal chance of being chosen.

2

EXERCISE 2C

1 Describe, with the aid of an example, the meaning of cluster sampling. Give one advantage and one disadvantage of cluster sampling compared to random sampling.

2 Explain, briefly, what you understand by cluster sampling. Under what circumstances would you advise its use?

Give one advantage and one disadvantage of cluster sampling compared to quota sampling.

3 The membership secretary of a football supporters club, which has 120 branches, wishes to contact a representative sample of members. She has a complete list of members classified by branches. She selects four branches at random and then ten members at random from each of the chosen branches.

(a) What name is given to this method of sampling?

(b) Would all members be equally likely to be included in the sample? Explain your answer.

(c) Under what (unlikely) circumstances would all members be equally likely to be included in the sample?

2.6 Systematic sampling

Sometimes items are selected on a regular pattern. For example every 100th vehicle from a production line could be tested or every 200th name on the electoral register could be chosen. In the case of the vehicles this ensures that the sample taken includes vehicles from throughout the day's production. It is not a random sample as, for example, it would be impossible for two consecutive vehicles to be both included in the sample.

Generally speaking systematic samples are perfectly satisfactory unless the pattern of the sampling follows a pattern in the population. For example if the first vehicle tested is always the 100th one produced each day it may be that the first few vehicles produced tend to be less satisfactory than the later ones and this would not be detected. Alternatively if it is known that every 100th vehicle is to be tested then extra care may be taken with these vehicles and so they are not typical of the whole population.

> In a systematic sample members of the population are chosen at regular intervals.

Quota, cluster and systematic samples are used to overcome the practical problems associated with random sampling but a random sample is preferable.

Worked example 2.4

The 985 pupils of Aberdashers Comprehensive School are listed in alphabetical order and numbered, consecutively, from 000 to 984. A sample of 20 pupils is required to take part in a survey on eating habits.

A number between 0 and 34 is selected at random. The corresponding pupil and every 50th pupil thereafter is included in the sample (i.e. if the first number selected was 9, the pupils numbered 009, 059, 109 … 909 and 959 would be selected).

(a) What is the name given to this method of sampling?

(b) Would all pupils be equally likely to be included in this sample? Explain your answer.

An alternative method of sampling is suggested. A number between 0 and 49 is selected at random. The corresponding pupil and every 50th pupil thereafter is included in the sample.

(c) (i) Would all pupils be equally likely to be included in this sample? Explain your answer.

(ii) Give a reason why neither of the two methods of sampling considered above would yield a random sample.

(iii) Give a reason, other than anything mentioned in your answers to parts **(c)(i)** or **(c)(ii)**, why the second method of sampling might be unsatisfactory in the given circumstances.

Solution

(a) Systematic sampling.

(b) No, pupils numbered 035 to 049 could not be included.

(c) (i) Yes, all pupils have a probability of $\frac{1}{50}$ of being included.

(ii) In neither case would it be possible for all sets of 50 pupils to form the sample. For example a brother and sister next to each other in the alphabetical list could never both be included in the same sample.

(iii) If the number selected at random was between 35 and 49 there would only be 19 pupils in the sample.

> Pupils numbered 000 to 034 have a probability of $\frac{1}{35}$ of being chosen.

> Also in the first method of sampling all pupils do not have the same probability of being chosen.

EXERCISE 2D

1 Describe with the aid of a simple example what you understand by systematic sampling.

2 An alphabetical list of the 2700 employees of a distribution company is available. Describe how a systematic sample of size 50 could be selected.

3 In order to estimate the mean number of books borrowed by library users a librarian decides to record the number of books borrowed by a sample of 40 members using the library. He selects a random integer, r, between 1 and 5. He includes in his sample the rth person to leave the library one morning and every 5th person after that until his sample of 40 is complete.

(a) What name is given to this method of sampling?

(b) Is each of the first 200 people leaving the library that morning equally likely to be included in the sample? Explain your answer.

(c) Explain why the sample does not constitute a random sample of the first 200 people leaving the library that morning.

(d) Comment on whether the sample will provide a useful estimate of the mean number of books borrowed by library users.

Worked example 2.5

On a particular day there are 2125 books on the shelves in the **fiction** section of a library.

(a) Describe how random numbers could be used to select a random sample of size 20 (without replacement) from the 2125 books. [A]

The number of times each book in the sample has been borrowed in the last year is found by counting the appropriate date stamps inside the front cover. The sample mean is used as an estimate of the mean annual number of times borrowed for all books belonging to the **fiction** section of the library.

(b) How is the estimate likely to be affected by the fact that:
(i) some books will not be on the shelves due to being in the possession of borrowers;
(ii) new books will have been on the shelves for less than a year? [A]

(c) Give a reason, other than those mentioned in **(b)**, why the sample will not be representative of **all** the books in the library. [A]

(d) Discuss, briefly, how the problems identified in **(b)** and **(c)** of estimating the mean number of times per year books in the library are borrowed could be overcome.

Solution

(a) Number books 0000 to 2124 either physically or by their position on the shelves,

Select four-digit random numbers

Ignore repeats and > 2124

Continue until 20 numbers obtained

Select corresponding books.

(b) **(i)** The books in possession of borrowers are likely to be the more popular books and so borrowed more frequently on average. Estimate from books on shelves likely to be too low.

(ii) New books on shelves for less than a year will tend to lead to an underestimate of the mean number of times borrowed.

(c) Non-fiction books not included.

(d) Library may have computerised records which make sampling unnecessary. If not it will at least have a catalogue of all books and a random sample could be selected from this. Books in the sample which are out on loan could be checked when they are returned. The data for new books could be scaled up according to how long they had been on the shelves.

Worked example 2.6

As part of an investigation of sampling methods, 150 plastic rods of varying lengths are placed on a desk by a teacher. A student is invited to view this population of rods. She is asked to estimate the mean length of the rods by choosing a sample of three which she thinks will have a mean length similar to that of the population (i.e. there is no attempt to select a random sample). The three selected rods are then measured and their mean length calculated. This process is repeated by each of the students in the class.

The following eight **mean** lengths, in cm, were obtained.

4.7   5.3   4.4   6.2   5.2   4.9   3.9   4.8

The mean length of the population of rods is 3.4 cm.

(a) Comment on this method of estimating the mean length of the population. [A]

(b) Use the eight mean lengths to estimate the mean and the standard deviation of the sample means. [A]

The teacher now asks a student to obtain a random sample of size three from the rod population.

(c) Describe how the student could use random number tables to obtain such a sample. [A]

2

Each student then independently obtains a random sample of size three from the population. The rods are measured and their mean lengths calculated as before. The eight **mean** lengths of the random samples are as follows:

4.6 1.9 7.2 1.5 2.2 2.8 3.7 1.4

(d) Use the eight mean lengths to estimate the mean and the standard deviation of the means of random samples of size three. [A]

(e) Compare the means of the non-random samples with those of the random samples. Your comments should include reference to the population mean. [A]

(f) Estimate the standard deviation of the population of rod lengths. [A]

(g) A sample of size 16 is taken from the population and the mean length is found to be 3.8 cm. It is claimed to be a random sample. State, giving a reason, whether or not this claim is reasonable.

Solution

(a) All eight sample means are above the population mean suggesting that the method is biased.

(b) 4.925, 0.680.

(c) Number rods 000 to 149
Select three-digit random numbers
Ignore 150 and over and repeats
Continue until three numbers obtained
Select corresponding rods.

(d) 3.16, 1.97

(e)

Random sample means	Non-random sample means
Some above population mean some below (unbiased)	All above population mean (biased)
Mean of means close to population mean.	Mean of means not close to population mean.
More variable than non-random.	Less variable than random

This requires knowledge from other areas of statistics. There will always be some questions on an examination paper which are not restricted to a single topic.

(f) For population standard deviation σ, the standard deviation of the mean of three randomly selected observations is

$\dfrac{\sigma}{\sqrt{3}}$. Hence σ is estimated by $\sqrt{3} \times 1.97 = 3.41$.

(g) Estimated standard deviation of mean of random sample of

size 16 is $\dfrac{3.41}{\sqrt{16}} = 0.85$.

Hence 3.8 cm is about 0.5 standard deviations above the mean. The claim is plausible.

The population mean is given as 3.4 cm.

EXERCISE 2E

1 The ages, in years, of the students in a statistics class are given below. The letters 'pt' following an age indicate that the student was attending the course on a part-time basis. The other ages are for full-time students.

19	20	23	21	21	20	20	19	19	20	19	24
20	19	20	22	21	25	20	33	19	19	19	20
49pt	24pt	36pt	27pt	33pt	26pt	38pt	43pt	24pt	41pt	30pt	27pt

(a) Describe how random numbers could be used to take a stratified sample of size 12 from this class.

(b) The mean age of a stratified sample of size 12 is $\bar{x}$ and the mean age of a random sample of size 12 is $\bar{y}$. Which of $\bar{x}$ and $\bar{y}$ would you choose as an estimate of the class mean age? Explain your answer.

2 A fast-food shop employs a student to conduct a survey into the eating habits of its customers. The student uses random number tables to choose a number, r, between 1 and 8. On a particular day she arrives when the shop opens and asks the rth customer entering the shop to complete a questionnaire. She then asks every 8th customer who enters the shop to complete the questionnaire until she has asked a total of 50 customers.

(a) Are all the first 400 customers entering the shop that day equally likely to be asked to complete the questionnaire?

(b) Are the chosen customers a random sample of the first 400 customers entering the shop that day? Give a reason for your answer.

(c) Discuss briefly whether the views expressed by these customers are likely to be representative of the views of all customers.

3 In a particular parliamentary constituency there are 64 000 names on the electoral register. Of the electors, 32 000 live in property rented from the local authority, 21 000 live in owner-occupied property and 11 000 live in other types of property.

The following methods are suggested for choosing a sample of electors in order to carry out an opinion survey.

A   Use a random process to select 128 names from the electoral register.

B   Use a random process to select one of the first 500 names on the electoral register. Using this as a starting point select every 500th name.

C   Select 64 names at random from the electors living in property rented from the local authority, 42 names at random from the electors living in owner-occupied

2

property and 22 names at random from those living in other types of property.

(a) For **each** of the methods **A**, **B** and **C**:

   (i) name the type of sampling method

   (ii) state whether all the names on the electoral register are equally likely to be included in the sample.   [A]

(b) State, giving a reason, whether method **C** will produce a random sample of the electors.   [A]

(c) State, briefly, the difference between a sample obtained by method **C** and a quota sample. What is the advantage of a quota sample compared to a sample obtained by method **C**?   [A]

(d) Compare the usefulness of sampling methods **A** and **C** if the questions to be asked concerned local authority housing policy.   [A]

(e) How would your answer to part **(d)** be changed (if at all) if the questions to be asked concerned attitudes to the monarchy? Explain your answer.   [A]

4 There are 28 houses in Mandela Road, 14 on each side. The houses on one side of the road have even numbers and those on the other side of the road have odd numbers.

A total of 63 residents of Mandela Road are on the electoral register.

A market researcher wishes to interview seven of these residents. He decides to choose a sample of seven houses using the following procedure:

Step 1 Toss a coin and choose the side of the road with odd numbered houses if it falls heads and the side of the road with even numbered houses if it falls tails.

Step 2 Toss the coin again and select the lowest numbered house on the chosen side of the road if it falls heads and the second lowest if it falls tails.

Step 3 Select alternate houses on the chosen side of the road starting from the house chosen in step 2. (For example, a tail followed by another tail would result in him selecting the houses numbered 4, 8, 12, 16, 20, 24 and 28.)

(a) (i) Would all houses in Mandela Road be equally likely to be included in the sample? Explain your answer.   [A]

   (ii) Would the sample be random? Give a reason for your answer.   [A]

The market researcher knocks at each selected house and asks the person who opens the door to answer a questionnaire. Assume that each person who opens the door is on the electoral register and is willing to answer the questionnaire.

(b) Give **two** reasons why the people who answer the questionnaire are not a random sample from the 63 residents on the electoral register. [A]

(c) Describe how random numbers could be used to select a random sample of size seven (without replacement) from the 63 residents on the electoral register. [A]

5 A small trade union has 23 000 members divided (unequally) into 12 branches. The executive committee decides to survey 200 members to seek their opinions of the services provided by the union. The following four suggestions were made as to how the sample should be chosen.

Suggestion A Select 5 of the 12 branches at random. The sample consists of 40 members selected at random from each of the 5 chosen branches.

Suggestion B The sample is made up of members from all branches. The members are selected at random from each branch, the number from each branch being proportional to the size of the branch membership.

Suggestion C As Suggestion B except that, instead of members being selected at random, each branch secretary is asked to select a sample in the most convenient way.

Suggestion D The members are numbered 00 000 to 22 999. Two hundred different 5-digit random numbers between 00 000 and 22 999 are taken from random number tables and the corresponding members chosen.

(a) State which of the suggestions would yield a quota sample. [A]

(b) For each of the other three suggestions state the type of sample described. [A]

(c) For each of the four suggestions state whether or not each member of the union would have an equal chance of being selected. [A]

(d) State, with a reason, which of the suggestions is best from

  (i) a statistical point of view

  (ii) a practical point of view.

Key point summary

I A **random sample** of size n is a sample selected in such a way that all possible samples of size n have an equal chance of being selected.   *p26*

2 A **stratified sample** requires prior knowledge to be used to divide the population into strata. Random samples are then taken from each of the strata, usually in proportion to the size of each of the strata.   *p26*

3 If relevant prior information about the population is available, a stratified sample using this information is preferable to a random sample.   *p26*

4 A **quota sample** is a stratified sample where the samples from each of the strata are chosen for convenience and there is no attempt at random sampling.   *p28*

5 A **cluster sample** uses randomly chosen clusters of the population. It is not a random sample but contains an element of randomness.   *p30*

6 In a **systematic sample** members of the population are chosen at regular intervals.   *p31*

7 Quota, cluster and systematic samples are used to overcome the practical problems associated with random sampling but a random sample is preferable.   *p32*

Test yourself	What to review
1 A list of the names of the 60 members of a cookery club is available. Describe how random numbers could be used to take a random sample of size 10.	*Section 2.2*
2 How could a systematic sample of size 10 be taken from the list in question **1**? How would you ensure that all members of the club had an equal chance of being included in such a sample?	*Section 2.6*
3 A firm employs 80 men and 140 women. Describe how random numbers could be used to take a stratified sample of size 22 from the employees. What difficulty would arise if the sample size required was 20? How would this difficulty be dealt with?	*Section 2.3*
4 How would a quota sample of size 22 be taken from the employees of the firm in question **4**?	*Section 2.4*

Test yourself (*continued*)	**What to review**
5 The members of the sample chosen in question **3** are asked how long they take to travel to work on a normal weekday. If the average time taken by the male and female employees is the same would a random sample be preferable to the sample chosen? Explain your answer.	*Section 2.3*
6 A journalist wishes to interview members of a political party which has branches in all parts of the country. A complete list of members is available. Explain to the journalist the advantages of choosing a cluster sample.	*Section 2.5*

Test yourself ANSWERS

1 Number names 00 to 59; select two-digit random numbers; ignore repeats and > 59; when 10 have been obtained choose corresponding names.

2 Choose a number between 00 and 05. Starting at this number choose every sixth name. If the original number is chosen by a random process all members of the club will have an equal chance of being selected.

3 Take a random sample of eight of the 80 men (as described in question **1**) and a random sample of 14 of the 140 women. If the total sample was of size 20 the samples from each of the strata could not be exactly proportional to the size of the strata. This could not be allowed for in the analysis.

4 Choose eight of the 80 men and 14 of the 140 women in any convenient way.

5 There would be no advantage in taking a stratified sample compared to a random sample. The only disadvantage would be that it is more complicated to take a stratified sample.

6 A random sample would be impractical because of the amount of travelling which would be involved. A cluster sample where a few branches are chosen at random and members of these branches interviewed would avoid bias, by retaining a random element, but involve far less travelling.

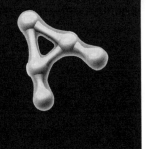

CHAPTER 3

Approximating distributions

Learning objectives

After studying this chapter, you should be able to:

- approximate a binomial distribution by a Poisson distribution
- approximate a binomial distribution by a normal distribution
- approximate a Poisson distribution by a normal distribution
- identify circumstances when each of these approximations is appropriate.

3.1 Approximating distributions

For a binomial distribution $P(X = r) = \binom{n}{r} p^r (1 - p)^{n-r}$, where n is the sample size and p is the probability of 'success'.

For the Poisson distribution $P(X = r) = e^{-\lambda} \dfrac{\lambda^r}{r!}$ where λ is the mean. Thus it is quicker to calculate Poisson probabilities than binomial probabilities. To tabulate binomial probabilities a range of values of n, p and r must be covered whereas for Poisson tables only values of λ and r are required. The Poisson tables are much more compact. The AQA tables contain several pages of binomial probabilities but there are still a lot of gaps. If you could approximate a binomial distribution which is not contained in the tables, say $n = 60$, $p = 0.005$, by a Poisson distribution you may be able to use tables of the Poisson distribution and at the least the amount of calculation would be reduced.

Similarly if you could approximate a binomial or a Poisson distribution, which were not contained in the tables, by a normal distribution, you could use the normal tables to obtain an approximate answer.

With the advent of powerful hand held calculators these reasons are becoming less important. It is possible to evaluate any Poisson or binomial distribution directly from these calculators. However normal approximations to the binomial and Poisson distributions are still required to calculate confidence intervals in Chapter 4.

Approximations are made:
– to reduce the amount of calculation
– to allow tables to be used
– to calculate confidence intervals.

3.2 Poisson approximation to the binomial distribution

Cars pass a point on a motorway independently at random at an average rate of three per minute. That is the number of cars passing the point follows a Poisson distribution with mean 3. The minute could be regarded as made up of 60 seconds. During each second either a car passes or no car passes. The probability of a car passing in a particular second is $\dfrac{3}{60} = 0.05$.

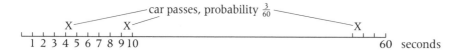

You now have a binomial distribution with $n = 60$ and $p = 0.05$. In fact the distribution is not quite binomial as it is possible that during a particular second more than one car will pass and so there are more than two possible outcomes. However the chance of this happening is small. If we divided the minute into even smaller intervals, say hundredths of a second, the chance of more than one car passing in a particular interval would be negligible. The more intervals taken, the smaller p becomes and so the more the binomial distribution resembles the Poisson distribution.

If n is large and p is small a binomial distribution may be approximated by a Poisson distribution.

Note. You will not be asked to justify this approximation in an examination, only to use it.

What is meant by large and small? Statistics is full of questions like this which have no precise answer. In fact it is much more important that p is small than that n is large. A commonly accepted rule of thumb is that n should be at least 50 and p should not be more than 0.1. It is probably best to stick to this rule although very good approximations can be made with much smaller values of n provided p is small enough.

There are no sharp dividing lines between good approximations and bad approximations.

Worked example 3.1

Andrea enters data into a computer in the form of binary digits. The probability that she will make an error is, independently, 0.002 for each digit entered. If she enters 700 digits find the probability that she will make three or fewer errors.

Solution

Binomial $n = 700$, $p = 0.002$. This will certainly not be contained in any table and deriving it from the binomial formula will involve an extremely length calculation. As n is large and p is small, a Poisson approximation may be used. The binomial distribution has mean np so a Poisson distribution with mean $700 \times 0.002 = 1.4$ will be used.

From tables P(3 or fewer) = 0.9463.

As this is an approximation it would be foolish to give the answer to four significant figures. Two or at most three is sufficient. In this case as n is so large and p is so small it is reasonable to give the answer as 0.946.

> The answer to four significant figures if the binomial distribution is used is 0.9465.

Worked example 3.2

The probability that a customer who enters a particular branch of a bank intends to open a new account is 0.0018 and is independent of the intentions of other customers.

Find the probability that:

(a) of 20 people who enter the bank, at least one intends to open a new account

(b) of 500 people who enter the bank three or more intend to open a new account.

Solution

(a) Binomial $n = 20$, $p = 0.0018$. Although p is small, n may not be big enough to use a Poisson approximation.
$$P(X \geqslant 1) = 1 - P(X = 0) = 1 - (1 - 0.0018)^{20} = 0.0354$$

(b) Binomial $n = 500$, $p = 0.0018$. n is large and p is small so approximate by a Poisson distribution with mean $500 \times 0.0018 = 0.9$.
From tables $P(X \geqslant 3) = 1 - P(X \leqslant 2) = 1 - 0.9371 = 0.063$.

> In this case p is so small that a Poisson approximation is adequate. However it is safer to stick to the rule that n should be at least 50.

EXERCISE 3A

1 The random variable, R, follows a binomial distribution with parameters n and p. Use a Poisson approximation if appropriate, or the exact binomial otherwise, to find the probability that

(a) $R \leqslant 6$ when $n = 90$ and $p = 0.1$

(b) $R < 4$ when $n = 110$ and $p = 0.05$

(c) $R = 2$ when $n = 150$ and $p = 0.02$

(d) $R \geqslant 5$ when $n = 120$ and $p = 0.01$

(e) $R \leqslant 2$ when $n = 15$ and $p = 0.15$

(f) $R > 16$ when $n = 50$ and $p = 0.35$

(g) $R = 6$ when $n = 100$ and $p = 0.1$.

2 The probability that a clerical worker calculating student fees makes an error is 0.004 for each calculation. Assuming the errors are independent, use a suitable approximation to find the probability that a random sample of 200 calculations will contain more than two errors.

3 The probability that a patient who agrees to take part in a clinical trial will leave the country before the trial is complete is 0.005.

(a) Use a suitable approximation to find the probability that of 1200 patients taking part in a trial, eight or more leave the country.

(b) What assumption have you had to make in order to carry out the calculation in part (a)?

4 A large batch of ballbearings contain 0.6% which are imperfect. They are packed at random into boxes of 200. Find the probability that a box will contain at least one imperfect ballbearing.

5 Yolande plays golf. The probability that she will succeed with a putt when the ball is less than 50 cm from the hole is 0.99. In the course of 12 games of golf she attempts 120 such putts. Assuming the outcomes are independent find the probability that she succeeds with at least 118 of them.

3.3 Normal approximation to the binomial distribution

In book S1 you met the central limit theorem. The second part of this states that the means of random samples from *any* distribution will be approximately normally distributed, provided the samples are large enough. In the binomial distribution samples of size n are taken from a population where the possible outcomes are 0 successes or 1 success.

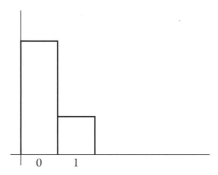

Sample 0, 0, 1, 0, 1, 0, 0, 0, 1, 0, 1, 1, 0, 0, 0,   mean $= \dfrac{5}{15}$

For large n the mean number of successes will be approximately normally distributed. The binomial distribution is the distribution of the total number of successes. The total is the mean multiplied by n. Multiplying by a constant will change the mean and standard deviation of a normal distribution but it will remain a normal distribution. Hence the central limit theorem justifies using the normal distribution as an approximation to the binomial distribution. This may be useful in circumstances when the Poisson approximation cannot be used and will certainly be useful for calculating confidence intervals in the next chapter.

> **Note.** You will not be asked to justify this approximation in an examination, only to use it.

 If n is large the binomial distribution may be approximated by a normal distribution

The question again arises what is meant by large? This depends to some extent on p. The larger p, the smaller n can be. A common rule of thumb is to use the approximation if n is at least 50 and np is at least 10.

> This rule assumes $p \leqslant 0.5$. If, say, probability of 'success' $= 0.9$, use $p = 1 - 0.9 = 0.1$ when applying this rule.

Worked example 3.3

The probability that a subject taking part in a clinical trial will withdraw before the trial is complete is 0.14 and is independent of whether other subjects withdraw or not. Use a suitable approximation to find the probability that of 600 subjects in the control group:

(a) exactly 85 withdraw

(b) 85 or more withdraw.

Solution

Binomial distribution $n = 600$ $p = 0.14$ $np = 600 \times 0.14 = 84$
Both n and np are easily large enough to use a normal approximation.

The binomial distribution has mean np and standard deviation $\sqrt{np(1 - p)}$ so the appropriate normal distribution will have mean 84 and standard deviation $\sqrt{600 \times 0.14 \times (1 - 0.14)} = 8.4994$.

(a) A problem now arises. The probability that an observation from a normal distribution equals 85 (or any other value) is zero. This problem has arisen because a discrete distribution is being approximated by a continuous distribution.

You may need to look back to S1, chapter 7 to revise the normal distribution.

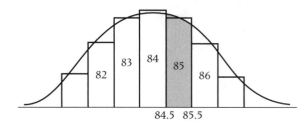

The best way forward is to approximate 85 (binomial) by the area between 84.5 and 85.5 (normal).

$$z_1 = \frac{(84.5 - 84)}{8.4994} = 0.0588$$

$$z_2 = \frac{(85.5 - 84)}{8.4994} = 0.1765$$

Interpolation has been used in the normal tables. This is desirable but a reasonably accurate answer can be obtained if the z values are rounded to two decimal places.

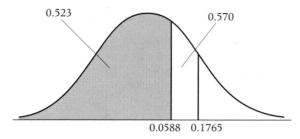

Estimated probability of 85 withdrawals $= 0.570 - 0.523 = 0.047$

Answer using the binomial distribution is 0.0463.

(b) Using the same argument as in part **(i)**, 85 or more (binomial) is approximated by >84.5 (normal). This is known as a continuity correction.

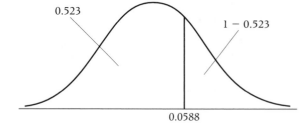

Estimated probability of 85 or more $= 1 - 0.523 = 0.477$

Answer using the binomial distribution is 0.471.

Worked example 3.4

Mushtaq organises a raffle at a fund raising evening for Oxfam. The probability that a person asked to buy a ticket will agree is 0.96. Find, using a normal approximation if appropriate, the probability that if Mushtaq asks:

(a) 60 people to buy tickets all will agree

(b) 190 people to buy tickets 180 or more will agree.

Solution

(a) Binomial $n = 60$. To consider whether an approximation is appropriate remember that p should be $\leqslant 0.5$ so you need to work in terms of people who do **not** agree to buy tickets. $p = 0.04$, $np = 2.4$. This is too small for a normal approximation. A Poisson approximation could be considered but it is quite straightforward to calculate this directly.

Probability all agree $= 0.96^{60} = 0.0864$

(b) Binomial $n = 190$, $p = 0.04$, $np = 7.6$. Since 7.6 is greater than 5 a normal approximation is appropriate.

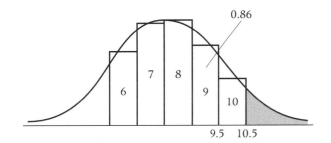

Mean 7.6 standard deviation $\sqrt{190 \times 0.04 \times 0.96} = 2.701$
180 or more agreeing $\rightarrow$ 10 or fewer **not** agreeing

$$z = \frac{(10.5 - 7.6)}{2.701} = 1.074$$

Estimated probability of 10 or fewer not agreeing $= 0.86$

EXERCISE 3B

1 The random variable, X, follows a binomial distribution with parameters n and p. Use a normal approximation if appropriate, or the exact binomial otherwise, to find the probability that

    (a) $X \leqslant 30$ when $n = 80$, $p = 0.3$

    (b) $X > 12$ when $n = 105$, $p = 0.15$

    (c) $X \leqslant 12$ when $n = 130$, $p = 0.12$

    (d) $X = 10$ when $n = 90$, $p = 0.1$

    (e) $6 \leqslant X \leqslant 14$ when $n = 200$, $p = 0.08$

    (f) $X \geqslant 7$ when $n = 30$, $p = 0.25$

    (g) $X \geqslant 49$ when $n = 50$, $p = 0.98$

    (h) $X = 16$ when $n = 70$, $p = 0.2$.

2 Traffic engineers observed that at a crossroads 35% of vehicles turned right. Use a normal approximation to find the probability of 100 vehicles reaching the crossroads

    (a) 30 or fewer turned right

    (b) exactly 35 turned right.

3 Boxes of chocolate are made up of chocolates randomly selected from large batches which contain 55% with hard centres and 45% with soft centres.

Small boxes contain 13 chocolates.

(a) Find the probability that a small box will contain
  (i) five or fewer with soft centres
  (ii) exactly five with soft centres.

Large boxes contain 60 chocolates.

(b) Find the probability that a large box will contain
  (i) 30 or more with soft centres
  (ii) between 25 and 30 (inclusive) with soft centres
  (iii) less than 30 with **hard** centres.

4 Ten percent of pint pots produced by a glass manufacturer contain flaws.

(a) A customer orders 12 pint pots. Find the probability that of the 12:

  (i) two or more will contain flaws

  (ii) exactly one will contain flaws.

(b) Another customer orders 120 pint pots. Find the probability that of these pint pots:

  (i) 20 or more will contain flaws

  (ii) less than 15 will contain flaws

  (iii) between 105 and 115 (inclusive) will be without flaws.

3.4 Normal approximation to the Poisson distribution

As well as justifying the use of the normal distribution as an approximation to the binomial distribution, the central limit theorem may also be used to justify the use of the normal distribution as an approximation to the Poisson distribution.

> A Poisson distribution with a large mean may be approximated by a normal distribution.

As before the definition of large is somewhat arbitrary but as rule of thumb the mean λ, should be at least 10.

Worked example 3.5

The number of letters delivered to a particular household follows a Poisson distribution with mean 25 per week.

Use a suitable approximation to find the probability that in a particular week:

(a) exactly 27 letters are delivered

(b) more than 27 letters are delivered.

Solution

The mean of 25 is sufficiently large to make a normal approximation. A Poisson distribution with mean λ, has standard deviation $\sqrt{\lambda}$ so the appropriate normal distribution will have mean 25 and standard deviation 5.

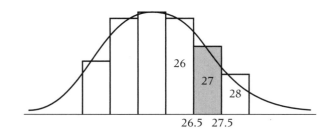

Again a discrete distribution is being approximated by a continuous distribution and so the probability of exactly 27 (Poisson) will be approximated by the probability of between 26.5 and 27.5 (normal).

(a) $z_1 = \dfrac{(26.5 - 25)}{5} = 0.3$

$z_2 = \dfrac{(27.5 - 25)}{5} = 0.5$

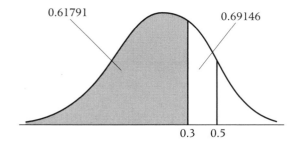

Probability of 27 letters in a
week $= 0.69146 - 0.61791 = 0.074$

Poisson gives 0.0708.

(b) Probability of more than 27 letters in a
week $= 1 - 0.69146 = 0.309$

Poisson gives 0.300.

Worked example 3.6

Telephone calls arrive at switchboard independently, at random with a constant average rate 8 per 15-minute interval. Using a normal approximation, if appropriate, find the probability of:

(a) 10 or more calls arriving in a 15-minute interval

(b) 40 or more calls arriving in an hour.

Solution

(a) A mean of 8 is a little low for a normal approximation, and in any case the answer can be found directly from the Poisson tables.

Probability of 10 or more calls in a 15-minute interval $= 1 - 0.7166$
$$= 0.283$$

(b) Number of calls in an hour will follow a Poisson distribution with mean $4 \times 8 = 32$. This is plenty large enough to use a normal approximation.

Mean 32, standard deviation $\sqrt{32} = 5.657$

$$z = \frac{(39.5 - 32)}{5.657} = 1.326$$

Probability of 40 or more $= 1 - 0.908 = 0.092$.

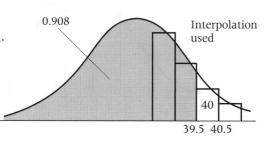

EXERCISE 3C

1 The random variable R follows a Poisson distribution with mean λ. Use a normal approximation, if appropriate, or the exact Poisson distribution otherwise, to find the probability that:

(a) $R \leqslant 20$ when $\lambda = 16$

(b) $R < 14$ when $\lambda = 22$

(c) $R > 46$ when $\lambda = 40$

(d) $110 \leqslant R \leqslant 120$ when $\lambda = 115$

(e) $R \geqslant 5$ when $\lambda = 2.6$

(f) $R = 18$ when $\lambda = 19$

(g) $R = 4$ when $\lambda = 3$

2 The number of customers arriving at a motorway cafe follows a Poisson distribution with mean 55 per 10-minute interval. Use a normal approximation to find the probability that in a particular 10-minute interval:

(a) more than 65 customers will arrive

(b) between 60 and 70 (inclusive) customers will arrive.

3 A company produces carpet tiles. The number of orders received for the tiles follows a Poisson distribution with a mean of 2.2 per working day. Using an approximation, if appropriate, find the probability that:

(a) no orders are received on a particular working day

(b) more than three orders are received on a particular working day

(c) a total of fewer than 35 orders are received in a 4-week (20-working day) period

(d) a total of more than 100 orders are received in an 8-week (40-working day) period.

4 Plants of a certain species are distributed over an area of heathland independently, at random, with a mean of 3.8 plants per square metre.

    (a) A biology student counts the number of plants in a 1 metre square. Find the probability of the square containing:

        (i) exactly three plants

        (ii) three or more plants.

    (b) The student now marks out a square of side 10 metres. Find the probability that this larger square contains:

        (i) 350 or more plants

        (ii) between 350 and 400 (inclusive) plants.

Worked example 3.7
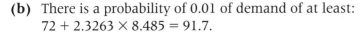

Demand for boxes of chocolates at a corner shop may be modelled by a Poisson distrubution with mean 2.4 per day. The owner wishes to obtain sufficient stock to last for the next 30 days. Use a suitable approximation to estimate:

(a) the probability that demand for boxes of chocolates during the 30-day period will exceed 80

(b) the number of boxes of chocolates which should be stocked in order that the probability of demand exceeding stock during the 30-day period is approximately 0.01.

Solution

Demand over 30 days is Poisson mean $30 \times 2.4 = 72$. Approximate by normal, mean 72, standard deviation $\sqrt{72} = 8.485$

(a) $z = \dfrac{(80.5 - 72)}{8.485} = 1.002.$

Probability demand exceeds 80 is $1 - 0.842 = 0.158$.

(b) There is a probability of 0.01 of demand of at least: $72 + 2.3263 \times 8.485 = 91.7.$

If 92 are stocked the probability of demand exceeding stock would be approximately 0.01.

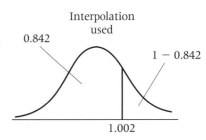

Interpolation used

0.842

$1 - 0.842$

1.002

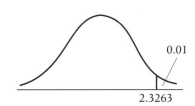

0.01

2.3263

Worked example 3.8

Meela starts a new job calculating housing benefit entitlements. At the end of her preliminary training the probability that her result is incorrect is 0.4 for each calculation.

(a) Find the probability of her obtaining less than five correct results from 10 calculations.

After a week's experience the probability of her result being incorrect has been reduced to 0.18 per calculation.

(b) In 80 calculations find:

    (i) the probability of there being at least nine incorrect

    (ii) the number of incorrect calculations which will be exceeded with a probability of approximately 0.01.

Meela has now been in the job for a year and the probability of her making an error is 0.03 for each calculation.

(c) In 80 calculations find:

    (i) the probability of there being at least nine incorrect,

    (ii) the number of incorrect calculations which would be exceeded with a probability of approximately 0.01.

Solution

(a) Binomial $n = 10$, $p = 0.4$ from tables $P(R < 5) = 0.633$

(b) Binomial $n = 80$, $p = 0.18$ $np = 14.4$

Approximate by normal distribution mean 14.4 standard deviation $\sqrt{80 \times 0.18 \times 0.82} = 3.436$

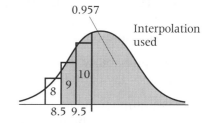

    (i) $z = \dfrac{(8.5 - 14.4)}{3.436} = -1.717.$

         Probability at least nine incorrect = 0.957

    (ii) Normal distribution will exceed $14.4 + 2.3263 \times 3.436$ $= 22.4$ with probability approximately 0.01. Since binomial is discrete 23 or more incorrect will have probability approximately 0.01.
Hence 22 will be exceeded with probability 0.01.

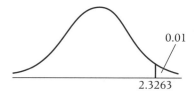

(c) Binomial $n = 80$, $p = 0.03$, $np = 2.4$.

     Approximate by Poisson distribution mean 2.4

    (i) from tables $P(R \geqslant 9) = 1 - 0.9991 = 0.0009$

    (ii) from tables $P(R \leqslant 6) = 0.9884.$

         Hence six exceeded with probability approximately 0.01.

EXERCISE 3D

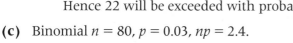

1 A technician looks after a number of automatic machines and has to make numerous minor adjustments. The necessity for these occurs at random at a constant average rate of eight per hour.

Using approximations if appropriate find the probability of:

(a) more than 70 adjustments being required in an 8-hour shift,

(b) between 60 and 70 (inclusive) adjustments being required in an 8-hour shift.

2 A golfer practises on a driving range. Her objective is to drive the ball to within 20 m of the flag.

   (a) On her first visit the probability of success with each drive is 0.3. If she drives ten balls find the probability of:

      (i) four or fewer success

      (ii) four or more successes.

   (b) Some weeks later the probability of success has increased to 0.46.
Find the probability of 120 or more successes in 250 drives.

   (c) A year later the probability of success has increased to 0.99. Find the probability of 4 or fewer failures in 120 drives.

3 In some circumstances the binomial distribution may be approximated by other distributions. In each of the following cases n is the number of bicycle journeys and p is the probability of the journey taking less than 1 hour. State whether a normal distribution, a Poisson distribution or neither would be an appropriate approximation.

   (a) $n = 110$, $p = 0.001$ **(b)** $n = 120$, $p = 0.45$

   (c) $n = 8$, $p = 0.14$ **(d)** $n = 130$, $p = 0.99$.

State why an approximation, where appropriate, may be useful.

4 A car hire firm has 120 cars available for hire. Daily demand for its cars follows a Poisson distribution with mean 104.

   (a) Use a suitable approximation to estimate the probability of the firm being unable to meet the demand on a particular day.

   (b) A student suggests that the calculation in **(a)** could be carried out using a binomial distribution with $n = 1040$ and $p = 0.1$ as an approximation. State whether:

      (i) this suggestion is correct

      (ii) this suggestion is sensible.

5 A pilot study suggests that the probability of a member of a particular trade union responding to a postal survey is 0.4.

   (a) Estimate the probability that at least 100 members will respond if

      (i) 270 are asked

      (ii) 290 are asked.

   (b) Without further calculation use your answers to part **(a)** to make a rough estimate of the number of members to ask in order to have a probability of about 0.95 that at least 100 members will respond.

6 Demand for replacement windscreens at a garage may be modelled by a Poisson distribution with mean 7 per day. Find:

    (a) the probability of demand on a particular day being less than 5

    (b) the demand which will be exceeded with a probability of about 0.05 on a particular day

    (c) the probability of demand being less than 30 during a particular 5-day period

    (d) the demand which will be exceeded with a probability of about 0.05 in a 5-day period.

7 When Roger plays Bagatelle the probability of him scoring more than 300 in a turn is 0.65.

    (a) Find the probability of him scoring more than 300 less than 60 times in 100 turns.

The probability of Roger scoring more than 500 in a turn is 0.04.

    (b) Find the probability of him scoring more than 500 at least once in 100 turns.

Key point summary

1 The purpose of making an approximation is: *p42*
   – to reduce the amount of calculation
   – to allow tables to be used where they otherwise could not
   – to calculate confidence intervals (see next chapter).

2 The binomial distribution may be approximated by   *p42*
the Poisson distribution if $n \geqslant 50$ and $p \leqslant 0.1$.

3 The conditions for the approximations given in key   *p42*
points **2**, **4** and **5** are rules of thumb. They are not
sharp dividing lines between good approximations and
bad approximations.

4 The binomial distribution may be approximated by   *p45*
the normal distribution if $n \geqslant 50$ and $np \geqslant 10$.

5 The Poisson distribution may be approximated by   *p48*
a normal distribution if $\lambda \geqslant 10$.

Test yourself	**What to review**
1 A binomial distribution has $n = 60$, $p = 0.02$. Use a suitable approximation to find $P(R > 1)$.	*Section 3.2*
2 A Poisson distribution has mean 32. Explain why the probability of 40 or more may be approximated by the probability of 39.5 or more from a normal distribution.	*Sections 3.3 and 3.4*
3 A binomial distribution has $n = 90$, $p = 0.4$. Use a suitable approximation to find $P(R < 30)$.	*Section 3.3*
4 It is suggested that a normal distribution with mean 174 and standard deviation 10.2 should be approximated by a binomial distribution with $n = 435$ and $p = 0.4$. Is this suggested approximation valid? Is it sensible?	*Section 3.3*
5 A Poisson distribution has a mean of 36. Use a suitable approximation to find $P(30 \leqslant R \leqslant 36)$.	*Section 3.4*

3

Test yourself ANSWERS

1 0.337

2 A discrete distribution is being approximated by a continuous distribution. 40 (discrete) must be approximated by 39.5 to 40.5 (continuous). Hence 40 or more must be approximated by 39.5 or more.

3 0.081.

4 Valid but not sensible because it is easier to deal with the normal distribution than with a binomial distribution with these parameters.

5 0.394.

CHAPTER 4

Confidence intervals

Learning objectives

After studying this chapter, you should be able to:

■ calculate an approximate confidence interval for the mean of a Poisson distribution
■ calculate an approximate confidence interval for a proportion
■ recognise circumstances in which the approximations are appropriate.

4.1 Confidence intervals

A supermarket manager wishes to estimate the mean number, λ, of customers entering the shop per minute on Monday mornings. He counts the number entering on 10 randomly selected minutes and calculates the **sample** mean as 12.4. However this is unlikely to be exactly the same as the **population** mean. If he had counted for a different 10 randomly selected minutes he would have found a different sample mean, say, 10.7. He can estimate the population mean, λ, but there is some uncertainty in his result. This uncertainty is expressed using a confidence interval. For example he could say that a 95% confidence interval for λ is

   12.4 ± 2.2

A 95% confidence interval means that 95% of confidence intervals calculated in this way will contain λ. Unfortunately this implies that 5% will not and there is no way of knowing whether this is one of the 95% which do contain λ or one of the 5% which do not. We can however say that it is much more likely to be one of the 95% which do contain λ.

λ remains constant but is unknown. Different samples will lead to different intervals. 95% of the intervals will contain λ.

> If a $100(1 - \alpha)\%$ confidence interval, for a population parameter λ, is calculated there is a probability of $1 - \alpha$ that it will contain λ.

4.2 Approximate confidence intervals for the Poisson distribution

It is possible to calculate an exact confidence interval for the mean of a Poisson distribution. The theory and necessary calculations are complex. However, if it is possible to approximate the Poisson distribution by a normal distribution, the problem is much more straightforward.

A traffic engineer counted 121 cars pass a point on a motorway during a 10-minute period. If the number of cars, X, passing the point follows a Poisson distribution with mean λ then X will be approximately normally distributed with mean λ and standard deviation $\sqrt{\lambda}$. The best estimate of λ available is 121 and so, X, can be approximated by a normal distribution with standard deviation $\sqrt{121} = 11$.

> This will be an adequate estimate of the standard deviation even if 121 is a poor estimate of the mean. For example if $\lambda = 144$, standard deviation $= 12$. 121 is a poor estimate of 144 but 11 is an adequate estimate of 12.

If an observation is taken from a standard normal distribution there is a probability of 0.95 that it will lie in the range ± 1.96.

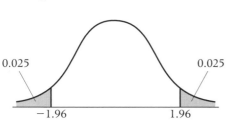

This implies that, for a normal distribution with mean λ and standard deviation σ, there is a probability of 0.95 that X will lie in the range $\lambda \pm 1.96\sigma$.

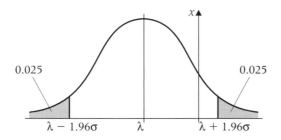

> The observed value x is known but λ is unknown.

It is not possible to calculate the interval $\lambda \pm 1.96\sigma$ because when calculating confidence intervals x is known but λ is unknown. However if x lies in the interval $\lambda \pm 1.96\sigma$ then the interval $x \pm 1.96\sigma$ will contain λ.

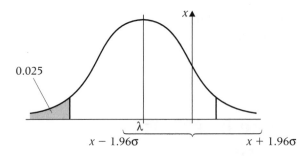

There is a probability of 0.95 that the interval $x \pm 1.96\sigma$ contains λ.

To return to the traffic engineer who counted 121 cars in a 10-minute period, an approximate 95% confidence interval for λ is

$$121 \pm 1.96 \times 11, \text{ i.e. } 121 \pm 21.6 \text{ or } 99 \text{ to } 143.$$

If the engineer feels that 95% is insufficient confidence then she can calculate 99% or 99.9% or any other percentage confidence interval. For example a 99% confidence interval would be

$$121 \pm 2.5758 \times 11, \text{ i.e. } 121 \pm 28.3, \text{ or } 93 \text{ to } 149$$

The percentage confidence has increased but the interval is wider. In fact all that is happening is that the same information is being displayed in a different way. The only way to improve the estimate is to collect more data.

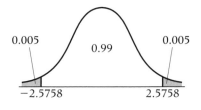

> The percentage confidence can be increased by widening the interval but 100% confidence will never be reached.

> If x is an observation from a Poisson distribution with mean λ then an approximate $100(1 - \alpha)\%$ confidence interval for λ is given by
> $$x \pm z_{\frac{\alpha}{2}}\sqrt{x}$$
> provided x is reasonably large, say >20.

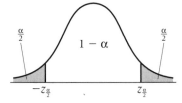

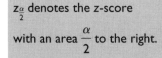

$z_{\frac{\alpha}{2}}$ denotes the z-score with an area $\dfrac{\alpha}{2}$ to the right.

Worked example 4.1

A shop sells 96 garden spades in a particular week. Assume that this can be regarded as a random sample from a Poisson distribution with mean λ.

(a) Calculate an approximate 90% confidence interval for λ.

(b) The shop manager claims that the shop sells an average of 100 spades per week. Comment on this claim.

(c) Calculate an approximate 90% confidence interval for the mean number of spades sold in a 4-week period.

(d) Give two reasons why the confidence intervals calculated in parts **(a)** and **(c)** are approximate and not exact.

Solution

(a) 90% confidence interval for λ is:

$96 \pm 1.6449 \times \sqrt{96}$, i.e. 96 ± 16.1
i.e. 80 to 112.

(b) 100 is comfortably within the confidence interval so there is no reason to doubt the shop manager's claim.

(c) Since the sum of four independent Poisson distributions with mean λ is a Poisson distribution with mean 4λ the confidence interval is found by simply multiplying the limits in part **(a)** by 4.

i.e. 320 to 448

Note. Since we only have data for 1 week, the confidence interval must first be calculated for 1 week and then scaled up (or down) as required.

(d) Approximate because
   (i) normal approximation to the Poisson distribution has been used
   (ii) the observed value of 96 is unlikely to be exactly equal to the mean, λ, and so the value $\sqrt{96}$ used for the standard deviation is only an approximation.

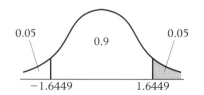

See S1, Chapter 6.

It is possible to allow for the continuity correction when calculating these intervals. This would make little difference and is beyond the syllabus.

4

Worked example 4.2

Cloth is woven in lengths of about 100 m. If a serious blemish is found in the cloth, a section containing the blemish is cut out and the two remaining parts stitched together. This is known as a 'string'. The number of 'strings' in lengths of cloth woven at a particular mill follows a Poisson distribution with mean m per length of cloth.

15 lengths of cloth are examined and the following numbers of strings found in each.

$$0 \quad 3 \quad 1 \quad 0 \quad 2 \quad 2 \quad 2 \quad 3 \quad 0 \quad 1 \quad 0 \quad 4 \quad 0 \quad 3 \quad 1$$

Calculate an approximate 95% confidence interval for m.

Solution

The total number of strings found in 15 lengths of cloth is 22. Since the number of strings in a length of cloth follows a Poisson distribution with mean m, the total number of strings in 15 lengths of cloth will follow a Poisson distribution with mean $15m$.

An approximate 95% confidence interval for $15m$ is

$22 \pm 1.96\sqrt{22}$ $\qquad$ 22 ± 9.19

An approximate 95% confidence interval for m is

$\dfrac{22}{15} \pm \dfrac{9.19}{15}$ $\qquad$ 1.47 ± 0.61 or 0.86 to 2.08

It is possible to carry out this calculation using the mean number of strings per cloth. Using the total is more straightforward and also makes it easier to check that a normal approximation is appropriate.

EXERCISE 4A

1 The number of telephone calls arriving at a switchboard follows a Poisson distribution. The switchboard receives 32 calls in a 10-minute interval.
Calculate an approximate:

(a) 95% confidence interval for the mean number of calls in a 10-minute period

(b) 99% confidence interval for the mean number of calls in a 10-minute period

(c) 90% confidence interval for the mean number of calls in a 10-minute period

(d) 95% confidence interval for the mean number of calls in a 1-minute period

(e) 99.9% confidence interval for the mean number of calls in 15-minute period.

2 An alternative technology centre sells wind-up radios. The demand occurs independently at random. During a particular week 29 radios were sold.

(a) Calculate an approximate 95% confidence interval for the mean number of radios sold each week.

(b) Give two reasons why the confidence interval you have calculated is approximate and not exact.

(c) Calculate an approximate 90% confidence interval for the mean number of radios sold in a 2-week period.

3 Applications for driving tests are received at a test centre at a mean rate of λ per day. The Poisson distribution provides an adequate model for the number of applications received.
Calculate an approximate:

(a) 95% confidence interval for λ if 23 applications are received on a randomly chosen day

(b) 95% confidence interval for λ if 59 applications are received in a 2-day period

(c) 90% confidence interval for λ if 128 applications are received in a 5-day period

(d) 99% confidence interval for λ if 24 applications are received in a 3-day period.

4 The Poisson distribution with mean 40 provides an adequate model for the number of package holidays sold by a travel agent in a week. In the first week following an advertising campaign the number of holidays sold was 48.

(a) Calculate an approximate 95% confidence interval for the mean number of holidays sold in a week after the advertising campaign.

(b) Does your calculation in part **(a)** show that the advertising campaign has been successful? Explain your answer.

(c) In the second week after the campaign 53 holidays were sold. Use all the data to calculate a 95% confidence interval for the mean number of holidays sold in a week.

(d) Does your calculation in part **(c)** affect your answer to part **(b)**?

(e) What assumptions have you had to make in order to answer this question?

4.3 Approximate confidence intervals for proportions

It is possible to calculate exact confidence intervals for the parameter p of a binomial distribution. As with the Poisson distribution, the theory and necessary calculations are complex but, if the binomial distribution can be approximated by a normal distribution, the process is fairly straightforward.

A newsagent observes that of 140 people who came into her shop one morning to buy a newspaper, 63 also bought other items. Provided it is reasonable to treat these 140 people as a random sample of all customers, 63 is a sample from a binomial distribution with $n = 140$.

This can be approximated by a normal distribution with mean $140p$ and standard deviation $\sqrt{140p(1-p)}$. An approximate 95% confidence interval for $140p$ is:

$$63 \pm 1.96\sqrt{140p(1-p)}$$

The best estimate that can be made of p is $\dfrac{63}{140} = 0.45$ and so

the standard deviation is estimated by $\sqrt{140 \times 0.45 \times (1 - 0.45)}$ $= 5.886$.

> As with the Poisson distribution this will be an adequate estimate of the standard deviation even if 0.45 is not a good estimate of p.

An approximate 95% confidence interval for $140p$ is:

$$63 \pm 1.96 \times 5.885 \quad 63 \pm 11.5$$

An approximate 95% confidence interval for, p, the proportion of customers who buy items in addition to newspapers is:

$$\frac{63}{140} \pm \frac{11.5}{140}, \qquad 0.45 \pm 0.082, \qquad \text{or } 0.368 \text{ to } 0.532.$$

> It is possible to allow for the continuity correction, but this is beyond the syllabus. As the intervals are only approximate it is acceptable to ignore it.

If r is an observation from a binomial distribution with parameters n, p then an approximate $100(1 - \alpha)\%$ confidence interval for p is given by

$$\hat{p} \pm z_{\frac{\alpha}{2}} \sqrt{\frac{\hat{p}(1 - \hat{p})}{n}}$$

provided r is reasonably large, say >20.

$\hat{p} = \dfrac{r}{n}$, the proportion of 'successes' in the sample,

and is an estimate of p.

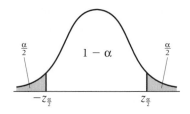

Worked example 4.3

In a large factory one or more reportable accidents occurred on 33 out of 200 working days.

(a) Calculate an approximate 90% confidence interval for the proportion of days on which one or more reportable accident occurs.

(b) The factory manager claims that reportable accidents occur on less than 10% of days. Comment on this claim.

(c) Give two reasons why the confidence interval calculated in part **(a)** is approximate and not exact.

Solution

(a) Binomial distribution, $n = 200$, p is estimated by

$\dfrac{33}{200} = 0.165$. Approximate 90% confidence interval for p is

> Assuming this can be regarded as a random sample of 200 days.

$$0.165 \pm 1.645 \times \sqrt{\frac{0.165 \times (1 - 0.165)}{200}}$$

$$0.165 \pm 0.043 \quad \text{or} \quad 0.122 \text{ to } 0.208$$

(b) 10% or 0.1 is below the lower limit of the confidence interval, so it is very unlikely that the proportion of days on which accidents occur is as low as the manager claims.

(c) Confidence interval is approximate because:

    (i) binomial distribution is approximated by normal distribution

    (ii) p is estimated from data and so the standard deviation used, $\sqrt{np(1 - p)}$, will not be exact.

Worked example 4.4 ───────────────────

A youth hostel warden observes that of the first 12 people booking into the hostel one afternoon four indicated they would take the evening meal provided and the others opted for self-catering.

(a) What problem would arise in calculating a confidence interval for the proportion of hostellers who would take the evening meal? How could this problem be overcome?

(b) Of 112 people arriving at the hostel 42 chose to take the evening meal. Assuming that the 112 hostellers can be treated as a random sample of all hostellers, calculate an approximate 99% confidence interval for the proportion of all hostellers who would choose to take the evening meal.

(c) If p is the proportion of all hostellers who would choose to take the evening meal, what is the probability that a 99% confidence interval does not contain p?

Solution

(a) Binomial $n = 12$, this value of n is too small to use a normal approximation. This could be overcome by increasing the size of n − that is by observing some more hostellers.

(b) Binomial $n = 112$, p is estimated by $\dfrac{42}{112} = 0.375$.

Approximate 99% confidence interval for p:

$$0.375 \pm 2.5758 \sqrt{\frac{0.375(1 - 0.375)}{112}}, \qquad 0.375 \pm 0.118$$

or 0.257 to 0.493.

(c) 0.01.

EXERCISE 4B ───────────────────

1 Random samples are selected from a large batch of components. A proportion, p, of the components is non-standard. The components in the sample are tested. Calculate an approximate

   (a) 95% confidence interval for p if 24 non-standard components are found in a sample of 520

   (b) 90% confidence interval for p if 48 non-standard components are found in a sample of 590

   (c) 99% confidence interval for p if 64 non-standard components are found in a sample of 156

   (d) 95% confidence interval for p if 24 non-standard components are found in a sample of 120.

2 Euan delivers free newspapers. Before taking the job he was told that only one household in 10 kept a dog. He observed that 24 out of 190 houses he delivered to had a dog.

   (a) Calculate an approximate 95% confidence interval for the proportion of households keeping a dog.

   (b) Comment on the claim that only one in 10 households keeps a dog.

   (c) What assumption did you have to make in order to answer parts **(a)** and **(b)**?

   (d) Give two reasons why the confidence interval you have calculated is approximate and not exact.

3 A traffic warden finds that of 82 parents who have arrived, by car, to collect children from a junior school 16 have parked illegally and dangerously on yellow lines. At a neighbouring school the number parking illegally is 22 out of 88. Assuming that in the long run the proportion of parents who park illegally is the same for both schools calculate an approximate 90% confidence interval for this proportion.

4 Eggs are classified as large, medium or small. Of 90 eggs laid at an organic farm on a particular day 34 were classified as large.

   (a) Calculate an approximate 95% confidence interval for the proportion of all eggs laid at this farm which are classified large.

   (b) What assumption have you had to make in order to calculate the confidence interval in **(a)**.

   (c) Comment on the farmer's claim that 40% of eggs laid on the farm are classified large.

   (d) What difficulty would have arisen in calculating the confidence interval if 86 of the 90 eggs had been classified large?

Worked example 4.5 _____

The number of faults in 100 m rolls of barbed wire may be assumed to have a Poisson distribution with mean λ. If a total of 94 faults is found in 25 such rolls:

(a) construct an approximate 99% confidence interval for λ

(b) what level of confidence (approximately) would be associated with an interval for λ of width 1 calculated from the given data?

(c) how many rolls, approximately, would it be necessary to examine in order to calculate a 90% confidence interval of width 1 for λ?

> Worked examples 4.5 and 4.6 are more difficult. Most examination questions will be easier than this.

Solution

(a) The mean number of faults on 25 rolls is 25λ.

Approximate 99% confidence interval for 25λ is:

$$94 \pm 2.5758\sqrt{94} \quad 94 \pm 24.98$$

i.e. 69 to 119.

Approximate 99% confidence interval for λ is 2.76 to 4.76.

(b) A confidence interval for 25λ is:

$$94 \pm z\sqrt{94}$$

A confidence interval for λ is:

$$\frac{94}{25} \pm \frac{z(\sqrt{94})}{25}$$

That is, it would be of width $\dfrac{2z(\sqrt{94})}{25}$.

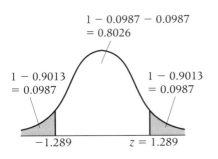

If it is of width 1 then $\dfrac{2z(\sqrt{94})}{25} = 1$,

$$z = 1.289.$$

Approximately 80% confidence would be associated with an interval of width 1.

> Interpolation has been used but this is not essential.

(c) If n rolls are examined and r faults are observed a 90% confidence interval for $n\lambda$ is:

$$r \pm 1.6449\sqrt{r}$$

A 90% confidence interval for λ is:

$$\frac{r}{n} \pm \frac{1.6449(\sqrt{r})}{n}$$

This is of width $2 \times \dfrac{1.6449(\sqrt{r})}{n}$.

If the interval is to be of width 1 then $\dfrac{3.2898(\sqrt{r})}{n} = 1$

$$\frac{(\sqrt{r})}{n} = 0.30397.$$

r is the number of faults observed in n rolls. Since we observed 94 faults in 25 rolls the best estimate we can

make of r is $\dfrac{94n}{25}$. Substituting this expression for r gives:

$$\frac{\sqrt{\left(\dfrac{94n}{25}\right)}}{n} = 0.30397$$

$$n = 40.7.$$

Hence it would be necessary to count the faults in about 41 rolls to calculate an approximate 90% confidence interval of width 1 for λ.

Worked example 4.6

In a large city, 98 adults, chosen at random, were asked when they last visited a dentist. The answers revealed that 24 of them had not visited a dentist for over 2 years.

(a) Calculate an approximate 95% confidence interval for the proportion of adults in the city who had not visited a dentist for over 2 years.

The dental practice that had commissioned the survey requested a confidence interval of width approximately 0.05.

(b) Find approximately the percentage confidence which would be associated with an interval of width 0.05 calculated from the data which has been obtained.

(c) Approximately how many adults should be asked in order to obtain an approximate 95% confidence interval of width 0.05?

(d) In parts **(b)** and **(c)** two different methods of reducing the width of the confidence interval in part **(a)** are investigated. Comment on these two methods.

Solution

(a) Estimate of proportion p is $\dfrac{24}{98} = 0.2449$

Approximate 95% confidence interval for p is:

$$0.2449 \pm 1.96 \sqrt{\frac{0.2449 \times (1 - 0.2449)}{98}}$$

$$0.245 \pm 0.085 \quad \text{or} \quad 0.160 \text{ to } 0.330$$

(b) Confidence interval is

$$0.2449 \pm z \sqrt{\frac{0.2449 \times 0.7551}{98}}$$

$$0.2449 \pm z \times 0.0434$$

The width of the interval is $2 \times z \times 0.0434 = 0.0868z$

For the interval to be of width 0.05, $z = 0.05/0.0868 = 0.576$

A 44% confidence interval would be of width approximately 0.05.

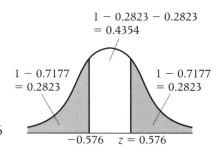

$1 - 0.2823 - 0.2823$
$= 0.4354$

$1 - 0.7177$
$= 0.2823$

$1 - 0.7177$
$= 0.2823$

$-0.576 \qquad z = 0.576$

> Interpolation has been used but this is not essential.

(c) An approximate 95% confidence interval for a sample of size n is:

$$\hat{p} \pm 1.96 \sqrt{\frac{\hat{p}(1 - \hat{p})}{n}} \text{ where } \hat{p} \text{ is an estimate of } p.$$

This is of width $2 \times 1.96 \sqrt{\dfrac{\hat{p}(1 - \hat{p})}{n}}$

If this is to be of width 0.05 and 0.2449 is substituted for $\hat{p}$

$$3.92\sqrt{\frac{0.2449 \times 0.7551}{n}} = 0.05$$

$$n = 1136.6.$$

It would be necessary to interview a random sample of approximately 1140 adults to obtain a 95% confidence interval of width 0.05.

(d) Reducing the width of the confidence interval by reducing the percentage confidence merely presents the same data in a different form. Increasing the sample size will genuinely improve the estimate, but in this case may have led to an inappropriately large sample for a local investigation.

EXERCISE 4C

1 Daisies are distributed independently, at random in a field with mean m per square metre. A randomly selected square metre is found to contain 22 daisies.

> Most examination questions will not be as difficult as Exercise 4C.

  (a) Calculate an approximate 95% confidence interval for m.

  (b) Calculate an approximate 99% confidence interval for the mean number of daisies to be found in an area of 10 square metres.

  (c) Approximately what percentage confidence would be associated with an interval for m of width:
    (i)   10 calculated from the data above
    (ii) 20 calculated from the data above?

  (d) Approximately what area would have to be examined for daisies in order to calculate:
    (i)   a 95% confidence interval of width 10 for m
    (ii) an 80% confidence interval of width 10 for m?

2 A sample of 100 jars of jam from a production line was examined and 33 of the jars were found to contain less than the amount of jam stated on the label.

  (a) Calculate an approximate 95% confidence interval for the proportion of jars which contain less than the stated amount.

  (b) Calculate an approximate 90% confidence interval for the proportion of jars which contain at least the stated amount.

  (c) What assumption was necessary to answer parts **(a)** and **(b)**?

  (d) Approximately what percentage confidence would be associated with an interval for the proportion of jars containing less than the amount of jam stated on the label of width:
    (i)   0.06 calculated from the data above
    (ii) 0.12 calculated from the data above?

(e) Approximately how many jars would it be necessary to test in order to calculate:

(i) a 95% confidence interval of width 0.06 for the proportion of jars which contain less than the stated amount

(ii) a 90% confidence interval of width 0.07 for the proportion of jars which contain less than the stated amount?

3 A traffic survey was carried out in order to estimate the mean number, μ, of vehicles passing a point on a motorway in a 1-minute interval during the rush hour. The same data is used to calculate both a 95% and a 99% confidence interval for μ. State the probability that:

(a) the 95% confidence interval does not contain μ

(b) the 99% confidence interval does contain μ

(c) the 99% confidence interval does contain μ but the 95% confidence interval does not

(d) the 95% confidence interval does contain μ but the 99% confidence interval does not

(e) the 99% confidence interval contains μ given that the 95% confidence interval does not

(f) the 95% confidence interval contains μ given that the 99% confidence interval does not

(g) the 95% confidence interval contains μ given that the 99% confidence interval contains μ.

Key point summary

1 If a $100(1 - \alpha)\%$ confidence interval, for a population parameter λ, is calculated there is a probability of $1 - \alpha$ that it will contain λ.   *p57*

2 The percentage confidence can be increased by widening the interval but 100% confidence will never be reached.   *p58*

3 If x is an observation from a Poisson distribution with mean λ then an approximate $100(1 - \alpha)\%$ confidence interval for λ is given by   *p58*

$$x \pm z_{\frac{\alpha}{2}}\sqrt{x}$$

provided x is reasonably large, say >20.

Key point summary *(continued)*

4 If r is an observation from a binomial distribution *p62*
with parameters n, p then an approximate $100(1 - \alpha)\%$
confidence interval for p is given by

$$\widehat{p} \pm z_{\frac{\alpha}{2}} \sqrt{\frac{\widehat{p}(1 - \widehat{p})}{n}}$$

provided r is reasonably large, say >20.

Test yourself	What to review
1 What is the probability that a 90% confidence interval for a population parameter γ does in fact contain γ?	*Section 4.1*
2 Is a 90% confidence interval wider, narrower or the same width as an 80% confidence interval?	*Section 4.2*
3 The number of three-wheeler cars passing a point on a main road follows a Poisson distribution. In a particular week four three-wheelers pass the point. What difficulty arises in using this data to calculate a confidence interval for the mean number of three-wheelers passing in a week?	*Section 4.2*
4 Out of 119 cars passing the point in question **3**, 32 had engines of less than 1000 cc. Calculate an approximate 95% confidence interval for the proportion of cars, with engines less than 1000 cc, passing this point.	*Section 4.3*
5 What assumption was it necessary to make in order to calculate the confidence interval in question **4**?	*Section 4.3*
6 The number of minor blemishes on mugs produced by a pottery follows a Poisson distribution with mean m per mug. Ten mugs are examined and the number of blemishes on each was: 0 4 2 3 2 1 3 2 0 5 Calculate an approximate 90% confidence interval for m.	*Section 4.2*

Test yourself ANSWERS

6 1.43 to 2.97.

5 That the sample could be regarded as a random sample of all cars passing the point.

4 0.189 to 0.349.

3 Estimated mean of the Poisson distribution is 4. This is too small for the normal distribution to give a good approximation.

2 Wider.

1 0.9.

CHAPTER 5

Interpretation of statistics

Questions on this section of the syllabus will be either based on a
table of secondary data or on a real application of statistics. There
are no specific techniques to learn although the syllabus does say
that questions may require knowledge of topics from S1 as well
as S2. This chapter consists of worked examples and exercises.

Worked example 5.1

The table below is copied from the *Annual Abstract of Statistics 2000*,
Office for National Statistics.

London Underground: receipts, operations and assets

	Unit	1988 /89	1989 /90	1990 /91	1991 /92	1992 /93	1993 /94	1994 /95	1995 /96	1996 /97	1997 /98	1998 /99
Receipts												
Passenger: total	£ million	432	462	531	559	589	637	718	765	797	899	977
Ordinary*	,,	228	262	295	307	322	350	396	430	449	510	547
Season tickets	,,	204	200	236	252	267	287	322	335	348	389	430
Traffic												
Passenger journeys: total	Million	815	765	775	751	728	735	764	784	772	832	866
Ordinary	,,	363	380	399	368	365	376	398	416	418	448	463
Season tickets	,,	452	385	376	383	363	359	366	368	354	384	403
Passenger kilometres	,,	6 292	6 016	6 164	5 895	5 758	5 814	6 051	6 337	6 153	6 479	6 716
Operations												
Loaded train kilometres	Million	50.5	50.1	52.4	52.5	52.5	52.7	54.8	57.2	58.6	62.1	61.2
Place kilometres	,,	43.6	43.0	44.9	45.2	45.3	45.6	49.4	51.6	52.2	55.5	54.8
Rolling stock												
Railway cars	Number	3 950	3 908	3 880	3 880	3 895	3 955	3 923	3 923	3 867	3 886	3 923
Seating capacity	Thousand	169.8	171.6	171.6	166.9	166.9	168.5	167.7	165.4	164.5	164.0	165.0
Permanent way and stations												
Route kilometres open for traffic	Kilometres	394	394	394	394	394	394	392	392	392	392	392
Stations	Number	246	245	246	246	246	245	245	245	245	245	245

*Includes one day travelcards and concessionary fares.

Source: Department of the Environment, Transport and the Regions.

(a) What were the London Underground receipts from season tickets in 1997/98?

(b) What were the average receipts per passenger kilometre in 1988/89 and in 1998/99?

(c) What was the average cost of a journey for passengers using ordinary (i.e. not season) tickets in 1988/89 and in 1998/99.

(d) Comment on your results in part **(c)** given that the United Kingdom Retail Price Index was 110.3 in December 1988 and 164.4 in December 1998.

Solution

(a) £389 million.

(b) Average receipts per passenger kilometre

1988/89 $\dfrac{432 \times 10^6}{6292 \times 10^6} = 0.0687$

   6.87p per passenger km

1998/99 $\dfrac{977 \times 10^6}{6716 \times 10^6} = 0.1455$

   14.55p per passenger km.

(c) Average cost of journey using ordinary ticket

1988/89 $\dfrac{228 \times 10^6}{363 \times 10^6} = 0.628$

   62.8p per journey

1998/99 $\dfrac{547 \times 10^6}{463 \times 10^6} = 1.18$

   £1.18 per journey

(d) Retail price index has risen from 110.3 to 164.4

$\dfrac{164.4}{110.3} = 1.49$. RPI has risen 49%

Average cost of journey has risen from 62.8p to £1.18,

$\dfrac{118}{62.8} = 1.88$. Has risen by 88%

Rise in average cost of journey far exceeds rise in RPI.

Be careful to include the units.

A ruler will help you to identify the required figure accurately.

5

Here 10^6 cancels but this will not always be the case. See Worked example 5.3**(d)**.

Worked example 5.2

The following table is copied from the *Annual Abstract of Statistics*, 1999, ONS.

United Kingdom airlines*
Operations and traffic on scheduled services: revenue traffic

	Unit	1988	1989	1990	1991	1992	1993	1994	1995	1996	1997	1998
Domestic services												
Aircraft stage flights:												
Number	Number	277 682	303 147	300 683	285 346	299 893	300 416	301 652	318 884	331 109	336 218	352 936
Average length	Kilometres	279.0	281.0	288.0	301.0	305.7	311.3	314.8	317.0	320.0	330.0	333.1
Aircraft-kilometres flown	Millions	77.4	85.2	86.5	86.0	91.6	93.5	94.9	101.1	105.8	111.0	
Passengers uplifted	,,	11.2	12.2	12.7	11.6	11.6	12.1	13.0	14.0	15.0	15.9	16.6
Seat-kilometres used	,,	4 381.1	4 767.6	5 020.8	4 663.7	4 728.2	4 933.8	5 334.0	5 753.6	6 204.3	6 745.7	6 947.5
Tonne-kilometres used:	Millions											
Passenger	,,	355.2	392.0	412.0	382.3	387.4	405.2	417.3	485.0	527.8	568.9	592.6
Freight	,,	10.1	8.9	8.7	6.7	6.6	5.6	6.3	6.9	7.4	6.1	6.0
Mail	,,	6.3	7.1	7.6	7.4	7.0	6.5	6.7	6.6	6.4	6.0	6.0
Total	,,	371.5	408.0	428.3	396.5	401.1	417.3	430.3	498.5	541.6	581.0	604.7
International services												
Aircraft stage flights:												
Number	Number	268 704	303 358	316 794	282 776	301 607	301 204	319 620	339 714	371 400	413 588	444 746
Average length	Kilometres	1 361	1 320	1 381	1 456	1 523	1 629	1 693	1 703	1 695	1 688	1 729
Aircraft-kilometres flown	Millions	365.8	400.5	437.4	411.7	459.5	490.8	541.4	578.8	629.5	698.2	768.8
Passengers uplifted	,,	20.2	22.9	25.7	22.9	26.5	28.0	30.9	33.5	36.1	40.4	45.1
Seat-kilometres used	,,	59 487.1	65 428.3	74 558.8	69 951.7	82 003.1	89 736.3	99.0	109.6	118.6	129.7	145.0
Tonne-kilometres used:	Millions											
Passenger	,,	5 514.6	6 212.9	7 053.6	6 625.6	7 747.8	8 500.1	9 352.3	10 636.4	11 661.9	12 718.2	14 162.3
Freight	,,	2 047.2	2 197.4	2 380.1	2 373.2	2 637.4	2 914.0	3 371.8	3 560.4	3 824.5	4 448.0	4 657.2
Mail	,,	172.9	155.2	161.1	175.2	154.0	135.0	140.5	144.4	169.5	166.3	171.7
Total	,,	7 734.8	8 565.6	9 594.7	9 174.0	10 539.3	11 549.1	12 864.6	14 391.2	15 655.9	17 332.5	18 991.2

*Includes services of British Airways and other UK private companies (including operations performed by Cathay Pacific Airways on their scheduled service London–Hong Kong from May 1981 until December 1984).

Source: Civil Aviation Authority.

(a) How many passengers were uplifted by United Kingdom airlines domestic services in 1993?

(b) Describe the trend exhibited by freight tonne-kilometres for:

  (i) domestic services

  (ii) international services.

  Compare the two series.

(c) The aircraft-kilometres flown for domestic services in 1998 has been obliterated from the table. Use other data from the table to calculate this figure. Indicate how you have calculated your answer.

Solution

(a) 12.1 million.

> Be careful to include 'million'.

(b) (i) There is a downward trend from 1988 to 1992. After that there is no clear trend, just apparently random variability.

(ii) There is a clear upward trend in this series. This is approximately linear although there is some indication that the rate of increase may be accelerating.

> Do not describe every small fluctuation in the series. Point out the main features.

The freight tonne-kilometres for international services is much larger than that for domestic services. It starts about 200 times as big and the differing trends lead to this ratio increasing rapidly.

(c) Aircraft-kilometres flown =
    (aircraft stage flights) × (average length)
For domestic services in 1998 = 352 936 × 333.1
                = 117 562 982
                = 117.6 million

5

Worked example 5.3

The following table contains data on student loans in the UK in 1990/91 and in 1998/99.

	1990/91			1998/99		
	Number of loans (thousand)	Total sum borrowed (£million)	Average size of loan (£)	Number of loans (thousand)	Total sum borrowed (£million)	Average size of loan (£)
Male						
Living at home	6	2	300	48	77	1600
Away from home (London)	14	6	430	34	77	2230
Away from home (Other)	85	33	390	237	444	1870
All male	105	41	390	320	598	1870
Female						
Living at home	3	1	300	45	72	
Away from home (London)	10	4	430	39	88	2230
Away from home (Other)	57	22	390	255	476	1870
All female	71	27	390	340	635	1870

(a) What was the total sum borrowed in 1998/99 by females living at home?

(b) If a pie chart of the number of loans taken out by females in 1990/91 was drawn

(i) what angle would be subtended at the centre by the sector representing females 'living at home'

(ii) what would be its radius, given that the radius of a similar pie-chart for 1998/99 was 4 cm?

(c) The total number of loans taken out by all males in 1998/99 is shown as 320 000. The total is made up of three categories 'living at home', 'away from home (London)' and 'away from home (other)'. The sum of the three numbers shown for these categories is $48 + 34 + 237 = 319$ thousand. Does this indicate that there must be an error in the data? Explain your answer.

(d) The average size of loan (£) for females living at home in 1998/99 is not shown. Calculate this value.

(e) Describe, briefly, the main features revealed by the table. Try to make four different points.

Solution

(a) £72 million.

(b) **(i)** There were 71 000 loans to females in 1990/91 of which 3000 were to females living at home. Angle subtended at the centre of a pie chart by this sector is

$$\left(\frac{3}{71}\right) \times 360 = 15.2 \text{ degrees.}$$

(ii) The pie chart for 1998/99 represents 340 000 loans.

The ratio of the areas is therefore $\dfrac{71}{340} = 0.2088$.

The radius of the 1990/91 chart is therefore:

$$\sqrt{0.2088} \times 4 = 1.8 \text{ cm.}$$

(c) This does not imply an error. It is probably simply due to rounding each of the categories to the nearest thousand.

(d) $\dfrac{£72\,000\,000}{45\,000} = £1600.$

(e) Big increase in number of loans from 1990/91 to 1998/99.

Big increase in average size of loans.

In 1990/91 many more males than females took loans. By 1998/99 more females than males took loans.

Size of loans for males and females nearly the same in all categories.

You may find other points to make.

Worked example 5.4

The table below is copied from the *Annual Abstract of Statistics*, 1999, ONS.

Motor vehicle production†
United Kingdom

	1988	1989	1990	1991	1992	1993	1994	1995	1996	1997	Number 1998
Motor vehicles											
SIC 1992, Class 34-10											
Passenger cars: total	1 226 835	1 299 082	1 295 610	1 236 900	1 291 880	1 375 524	1 466 823	1 532 084	1 686 134	1 698 001	1 748 258
1000 cc and under	129 446	133 135	93 039	15 918	22 037	98 034	98 178	95 198	108 645	119 894	112 044
Over 1000 cc but not over 1600 cc	764 289	716 784	809 219	496 822	793 307	709 615	729 397	814 873	845 084	829 086	814 595
Over 1600 cc but not over 2800 cc	260 231	375 309	325 116	193 972	437 951	515 487	573 357	528 444	635 861	653 154	720 556
Over 2800 cc	72 869	73 854	68 236	26 001	38 585	52 388	65 891	93 569	96 544	95 881	101 063
Commercial vehicles: total	317 343	326 590	270 346	217 141	248 453	193 467	227 815	233 001	238 314	237 703	227 379
Of which:											
Light commercial vehicles	250 053	267 135	230 510	105 633	216 477	171 141	197 285	199 346	205 372	210 942	203 629
Trucks: Under 7.5 tonnes	19 732	17 687	10 515	5 379	9 558	4 755	8 154	9 523	9 812	6 254	5 006
Over 7.5 tonnes	24 887	21 083	13 674	7 673	11 113	8 269	10 016	11 717	9 229	7 930	7 002
Motive units for articulated vehicles	6 171	5 827	3 327	1 444	2 788	2 283	2 794	3 476	208	2 573	2 492
Buses, coaches and mini buses	16 500	14 858	12 320	5 593	8 517	7 019	9 566	8 939	9 254	10 004	9 250

†Figures relate to periods of 52 weeks (53 weeks in 1988 and 1993).

Source: Office for National Statistics: 01633 812963.

(a) How many trucks over 7.5 tonnes were produced in the United Kingdom in 1995?

(b) How many motor vehicles in total were produced in the United Kingdom in 1996?

(c) The total number of passenger cars produced in 1998 is shown as 1 748 258. Comment on the number of significant figures given.

(d) Explain why the figures given for 1991 must contain some errors.

(e) The footnote suggests that the figures for 1988 and 1993 are compiled on a slightly different basis than the other years. Suggest a method of modifying the figures for 1988 and 1993 to give a fairer comparison.

(f) Describe, briefly, the overall trend shown by the total number of passenger cars produced in the United Kingdom and compare this with the trend shown by the total number of commercial vehicles produced.

Solution

(a) 11 717.

(b) 1 686 134 passenger cars + 238 314 commercial
vehicles = 1 924 448 motor vehicles.

(c) The seven significant figures given are excessive. It is highly
unlikely that this number could be correct to 7 sf or that
anyone would require this figure to such a level of accuracy.

(d) In 1991 (unlike other years) the totals are not equal to the
sum of the constituent parts. For example for passenger
cars:

$$1\,236\,900 \neq 15\,918 + 496\,822 + 193\,972 + 26\,001.$$

(e) Figures for 1988 and 1993 could be multiplied by $\dfrac{52}{53}$.

(f) Total passenger car production shows a fairly steady upward
trend apart from a dip around 1990 to 1992.

Commercial vehicle production fell sharply but erratically
from 1989 to 1993 but was fairly constant at around
230 000 from 1994 to 1998.

Worked example 5.5

The following information refers to legislation for pre-packaged
goods. It is extracted from the manual of 'Practical Guidance for
Inspectors'.

Package – a container, together with the predetermined
quantity of goods it contains, made up in the absence of the
purchaser in such a way that none of the goods can be removed
without opening the container.

Nominal quantity, Q_n – the quantity marked on the container.

Tolerable negative error, *TNE* – the negative error in relation
to a particular nominal quantity as defined by the table below.

Tolerance limit, T_1 – the nominal quantity minus the tolerable
negative error

$$T_1 = Q_n - TNE.$$

Absolute tolerance limit, T_2 – the nominal quantity minus
twice the tolerable negative error

$$T_2 = Q_n - 2TNE.$$

Non-standard package – a package whose contents are less
than the tolerance limit, T_1.

Inadequate package – a package whose contents are less than
the absolute tolerance limit, T_2.

Tolerable negative errors (weight or volume)

Nominal quantity, Q_n (g or ml)	Tolerable negative error, TNE	
	% of Q_n	g or ml
5–50	9	–
50–100	–	4.5
100–200	4.5	–
200–300	–	9
300–500	3	–
500–1000	–	15
1000–10 000	1.5	–
10 000–15 000	–	150
Above 15 000	1	–

Thus, for tins of lentils with nominal quantity 400 g, the *TNE* is 12 g, and a non-standard tin is one containing less than 388 g.

Packets of peppermint tea have nominal quantity 40 g.

(a) Find

    (i) the tolerable negative error

    (ii) the tolerance limit

    (iii) the absolute tolerance limit.

(b) If the contents of the packets are normally distributed with mean 40.5 g and standard deviation 2.5 g, find the proportion of packets which are:

    (i) non-standard

    (ii) inadequate.

(c) Legislation requires that:
- the average contents of packages must be at least Q_n
- not more than 2.5% of packages may be non-standard
- not more than 0.1% of packages may be inadequate.

    (i) Comment on the packets of peppermint tea in relation to these requirements.

    (ii) Suggest a possible reason why it might be easier to meet these requirements for packets of tea with nominal quantity 50 g than for packets of tea with nominal quantity 90 g.

Solution

(a) **(i)** $TNE = \dfrac{9}{100} \times 40 = 3.6 \text{ g}$

    (ii) $T_1 = 40 - 3.6 = 36.4 \text{ g}$

    (iii) $T_2 = 40 - 2 \times 3.6 = 32.8 \text{ g}$.

(b) (i) $z = \dfrac{36.4 - 40.5}{2.5} = -1.64$

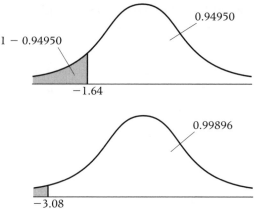

Proportion non-standard
$= 1 - 0.94950 = 0.0505$.

(ii) $z = \dfrac{32.8 - 40.5}{2.5} = -3.08$

Proportion inadequate
$= 1 - 0.99896 = 0.00104$.

(c) (i) Condition I met.
Condition II not met.
Condition III not met but difference is negligible.
To meet condition II, either mean must be increased or standard deviation reduced.

(ii) *TNE* has same magnitude for $Q_n = 50$ and $Q_n = 90$.
However it is 9% for $Q_n = 50$ and only 5% for $Q_n = 90$.
Probably easier to ensure 97.5% contain at least 91% of Q_n than 97.5% contain at least 95% of Q_n.
Similarly for inadequate.

EXERCISE 5A

1

GB cinema exhibitor statistics

	Sites	Screens	Total no. of admissions	Gross box office takings	Amount paid out for films	Revenue per admission
	(number)	(number)	(number)	(£million)	(£million)	(£)
1988	495	1 117	75.2	142.2	55.1	1.89
1989	481	1 177	82.9	169.5	65.4	2.04
1990	496	1 331	78.6	187.7	69.5	2.39
1991	537	1 544	88.9	229.7	78.5	2.58
1992	480	1 547	89.4	243.7	81.9	2.73
1993	495	1 591	99.3	271.3	95.9	2.73
1994	505	1 619	105.9	293.5	104.6	
1995	475	1 620	96.9	286.5	96.3	2.96
1996	483	1 738	118.7	373.5	144.9	3.15
1997	504	1 886	128.2	443.2	165.9	3.46
1998	481	1 975	123.4	449.5	162.6	3.64

Source: Office for National Statistics.

(a) How many admissions to GB cinemas were there in 1994?

(b) Describe the trend in the number of cinema:

(i) sites **(ii)** screens.

Extend your answer to include comments on the number of screens per site.

(c) The figure for revenue per admission in 1994 has been obliterated. Use the data available to calculate this figure.

(d) Comment on the revenue per admission given that the Retail Price Index was 106.6 in June 1988 and 163.4 in June 1998.

2 Landings of fish by United Kingdom vessels: live weight and value into United Kingdom

	Quantity (thousand tonnes)							Value (£ thousand)					
	1992	1993	1994	1995	1996	1997		1992	1993	1994	1995	1996	1997
Brill	0.5	0.4	0.5	0.5	0.5	0.5		1 629	1 502	1 927	2 201	2 639	2 244
Catfish	1.9	1.7	1.4	1.0	1.1	1.0		1 878	1 867	1 669	1 474	1 449	1 152
Cod	62.2	64.8	65.7	74.4	75.7	71.0		71 570	64 915	65 090	65 772	69 752	67 641
Dogfish	14.1	10.4	8.2	10.9	9.7	8.6		10 032	7 327	6 291	7 684	7 002	5 586
Haddock	53.9	87.1	92.9	85.3	89.1	82.6		40 071	54 710	61 017	54 734	54 333	44 778
Hake	4.6	4.3	3.2	3.1	2.8	2.7		10 757	11 535	8 363	7 361	7 346	6 375
Lemon Soles	5.6	5.0	4.8	4.6	5.1	5.2		11 195	11 073	11 931	10 115	11 897	11 786
Ling	6.1	7.7	8.1	9.8	9.2	9.4		4 901	5 210	5 832	7 483	6 945	6 568
Megrims	4.1	4.3	4.7	5.1	6.0	6.0		8 309	8 543	9 210	9 227	10 430	9 467
Monks or Anglers	15.0	16.7	17.7	22.2	29.7	25.6		30 910	32 271	34 328	39 451	51 468	45 947
Plaice	24.1	19.4	17.5	15.5	12.5	12.9		24 281	21 499	19 937	17 573	16 013	15 484
Pollack (Lythe)	3.1	3.4	2.9	3.3	2.9	3.0		3 039	3 074	2 564	2 935	2 563	2 491
Saithe	13.0	12.2	12.5	12.8	13.3	12.3		5 707	4 671	4 875	5 625	5 674	4 894
Sand Eels	4.2	2.9	8.7	7.3	9.3	14.5		185	124	424	398	505	815
Skates and Rays	7.9	7.3	7.3	7.6	8.3	7.2		5 884	5 899	6 322	6 543	7 451	6 215
Soles	2.9	2.6	2.8	2.8	2.5	2.3		13 161	12 795	13 454	14 200	14 068	14 800
Turbot	0.8	0.9	1.0	0.8	0.8	0.7		3 874	4 521	5 238	4 667	5 006	4 263
Whiting	44.2	45.6	41.8	39.8	37.3	34.5		21 680	20 748	20 135	18 918	18 962	15 939
Whiting, Blue	6.5	2.3	1.9	3.4	3.5	12.4		333	137	90	180	190	693
Whitches	2.1	2.0	2.1	2.3	2.3	2.3		2 459	2 305	2 867	2 915	3 068	2 263
Other Demersal	8.6	8.9	8.9	12.9	13.3	13.2		7 034	8 730	10 709	17 028	16 614	16 000

Source: Office for National Statistics.

(a) What quantity of dogfish was landed in the United Kingdom in 1996?

(b) What was the value of the dogfish landed in the United Kingdom in 1996?

(c) Describe, briefly, the trend in the quantity of:

(i) plaice landed **(ii)** sand eels landed.

(d) Calculate the average value of a tonne of haddock landed in:

(i) 1992 **(ii)** 1997.

(e) Calculate the average value of a tonne of brill landed in:

(i) 1992 **(ii)** 1997.

(f) Comment on the likely accuracy of your answers to part **(e)**.

(g) Comment on your answers to parts **(d)** and **(e)** given that the Retail Price Index in June 1992 was 139.3 and in June 1997 was 157.5.

3 The following table is copied from the Digest of Environmental Statistics No 20, 1998.

Number of councils* and number of sites† participating in the bottle bank scheme: 1977–1996

Great Britain

	Number of councils	Number of sites
1977	6	17
1978	20	76
1979	33	143
1980	119	433
1981	158	614
1982	232	1 230
1983	286	1 758
1984	316	2 144
1985	341	2 470
1986	341	2 842
1987	389	3 138
1988	416	3 653
1989	431	4 330
1990	444	5 842
1991	448	7 155
1992	456	8 703
1993	457	10 965
1994	458	12 858
1995	459	14 300
1996	431‡	15 609

Source: British Glass Manufacturers Confederation.

* Includes Isle of Man, Guernsey and Isles of Scilly.
† Best estimate, as exact number of sites is not known.
‡ The lower figure is entirely the result of local government reorganisation, which has resulted in a smaller number of councils.

(a) Describe, briefly, the trend in the number of sites participating in bottle bank schemes.

(b) Describe, briefly, the trend in the number of councils participating in bottle bank schemes.

(c) Why would it be misleading to include the 1996 figure in a time series graph showing the number of councils participating in bottle bank schemes?

(d) From 1979 to 1985 there was a rapid increase in the number of councils participating. Why would it be impossible for this rate of increase to continue until 1995?

(e) Find the average number of sites per participating council in 1977, 1982, 1988 and 1995. Comment.

4 The following table is extracted from 'Health in England 1998: investigating the links between social inequalities and health'.

Cigarette smoking status by gross household income and sex
Adults aged 16 and over

Cigarette smoking status	Current cigarette smoker (Observed %*)	Ex-cigarette smoker (Observed %*)	Never smoked (Observed %*)	Base = 100%
Gross household income				
Men				
£20 000 or more	26	28	46	1093
£10 000–£19 999	31	30	39	683
£5 000–£9 999	31	40	29	440
Under £5 000	48	29	22	197
Total	29	30	40	2413
Women				
£20 000 or more	21	20	60	982
£10 000–£19 999	27	20	52	726
£5 000–£9 999	28	22	49	740
Under £5 000	32	22	46	545
Total	25	21	54	2993

* The percentages refer to the proportions of the subgroups reporting each smoking status, so do not add to 100.

(a) According to the table, what percentage of men from households with gross income between £10 000 and £19 999 are ex-cigarette smokers?

(b) What percentage of women from households with gross income between £5 000 and £9 999 are ex-cigarette smokers?

(c) Describe the main feature of the data for women who have never smoked.

(d) Describe the main feature of the data for men who have never smoked and compare this data with the data for women who have never smoked.

(e) The final column shows that the data for men from households with income under £5000 is based on a sample of 197. Calculate an approximate 95% confidence interval for the proportion of ex-cigarette smokers in this group. What assumption was it necessary to make?

(f) Bearing in mind your answer to part **(e)**, compare the proportion of ex-cigarette smokers among men from households with income under £5000 and men from households with income £20 000 or more.

5

5 The following table is copied from the Annual Abstract of Statistics 1999.

Electricity: fuel used in generation – United Kingdom

Million tonnes of oil equivalent

	1988	1989	1990	1991	1992	1993	1994	1995	1996	1997	1998
All generating companies: total fuels	79.4	80.2	77.4	78.3	78.0	76.8	75.2	76.5	78.4	77.9	79.5
Coal	53.1	51.6	50.0	50.0	46.9	39.6	37.1	36.1	33.4	28.9	30.0
Oil	7.1	7.1	8.4	7.6	8.1	5.8	4.1	3.6	3.5		1.4
Gas	1.0	0.5	0.6	0.6	1.5	7.0	9.9	12.5	16.4	20.9	22.2
Nuclear	15.7	17.7	16.3	17.4	18.5	21.6	21.2	21.3	22.2	23.0	23.3
Hydro (natural flow)	0.4	0.4	0.4	0.4	0.5	0.4	0.4	0.5	0.3	0.4	0.4
Other fuels used by UK companies	1.0	1.7	1.8	0.9	1.1	1.0	1.1	1.2	1.2	1.4	1.2
Net imports	1.1	1.1	1.0	1.4	1.4	1.4	1.5	1.4	1.4	1.4	1.1

Source: Department of Trade and Industry.

(a) How much oil was used in electricity generation in the United Kingdom in 1994?

(b) Describe the trend in the use for electricity generation of:

(i) coal   **(ii)** gas   **(iii)** nuclear.

(c) If a pie chart was drawn showing the fuels used in electricity generation in the United Kingdom in 1998, what would be the angle subtended at the centre by the sector representing oil?

How would this sector compare with the sector representing oil in a similar pie-chart drawn for 1988?

(d) The table shows that the use of hydro (natural flow) was the same in each year from 1988 to 1991 and then increased by 25% in 1992. Comment on this.

(e) How much oil was used in electricity generation in the United Kingdom in 1997?

6 This question relates to the legislation for prepackaged goods which is described in Worked example 5.5 on pages 76 and 77. Jars of coffee granules have a nominal quantity of 100 g.

(a) Find:

(i) the tolerable negative error

(ii) the tolerance limit

(iii) the absolute tolerance limit.

(b) If the contents of the jars are normally distributed with mean 99 g and standard deviation 2 g find the proportion of packets which are:

(i) non-standard   **(ii)** inadequate.

(c) Comment on the jars of coffee in relation to the legal requirements.

(d) Packets of tea have nominal contents 75 g. They have mean contents of 76 g with a standard deviation of 4 g. Comment on these packets in relation to the legal requirements.

Time allowed 1 hour 15 minutes

Answer **all** questions

1 The number of orders a light engineering company receives
may be modelled by a Poisson distribution with mean λ per
week. The number of orders received last week was 42.

(a) Calculate an approximate 95% confidence interval
for λ. (*5 marks*)

(b) Comment on the sales manager's claim that the average
number of orders received per week is 50. (*2 marks*)

2 **Parliamentary elections**

Thousands

	25 Oct 1951	26 May 1955	8 Oct 1959	15 Oct 1964	31 Mar 1966	18 June 1970*	28 Feb 1974	10 Oct 1974	3 May 1979	9 June 1983	11 June 1987	9 April 1992	1 May 1997‡
United Kingdom													
Number of electors	34 919	34 852	35 397	35 894	35 957	39 615	40 256	40 256	41 573	42 704	43 666	43 719	43 830
Average-electors per seat	55.9	55.3	56.2	57.0	57.1	62.9	63.4	63.4	65.5	66.7	67.2	67.2	
Number of valid votes counted	28 597	26 760	27 863	27 657	27 265	28 345	31 340	29 189	31 221	30 671	32 530	33 551	31 286
As percentage of electorate	*81.9*	*76.8*	*78.7*	*77.1*	*75.8*	*71.5*	*77.9*	*72.5*	*75.1*	*71.8*	*74.5*	*76.7*	*71.4*
Number of Members of													
Parliament elected:	625	630	630	630	630	630	635	635	635	650	650	651	659
Conservative	320	344	364	303	253	330	296	276	339	396	375	336	
Labour	295	277	258	317	363	287	301	319	268	209	229	271	418
Liberal Democrat	6	6	6	9	12	6	14	13	11	17	17	20	46
Social Democratic Party	–	–	–	–	–	–	–	–	–	6	5	–	–
Scottish National Party	–	–	–	–	–	1	7	11	2	2	3	3	6
Plaid Cymru	–	–	–	–	–	–	2	3	2	2	3	4	4
Other†	4	3	2	1	2	6	15	13	13	18	18	17	20

* The Representation of the People Act 1969 lowered the minimum voting age from 21 to 18 years with effect from 16 February 1970.

† Including the Speaker.

‡ Provisional.   *Source: Home Office.*

(a) How many valid votes were counted at the election on
9th June 1983? (*2 marks*)

(b) Some of the figures in the last column have been obliterated.

    (i) How many Conservative members of parliament were elected on 1st May 1997?

    (ii) What was the average number of electors per seat at the election on 1st May 1997?   (*4 marks*)

(c) Describe, briefly, the trend of

    (i) 'Number of electors'

    (ii) 'Number of valid votes counted.'

    Compare these two trends.   (*6 marks*)

3 A supermarket employs Agyeman to conduct a survey into customer satisfaction. On a particular day Agyeman chooses a random number, r, between 1 and 12. He then asks the rth customer and every 12th customer thereafter to complete a questionnaire until he has asked a total of 50 customers.

(a) Are all the first 600 customers entering the supermarket that day equally likely to be included in the sample? Explain your answer.   (*3 marks*)

(b) Is the chosen sample a random sample of the first 600 customers entering the supermarket? Explain your answer.   (*3 marks*)

(c) The total number of customers entering the supermarket that day was 1912. If Agyeman had asked the rth customer (selected as before) and thereafter every 100th customer entering the supermarket would all customers entering the supermarket that day be equally likely to be included in this sample? Explain your answer.   (*3 marks*)

(d) Discuss briefly whether the views expressed by the customers selected by either method are likely to be representative of the views of all customers.   (*3 marks*)

4 The following table shows the expenditure on sea travel, in £ million, by UK households. The figures have been adjusted to constant 1995 prices. These should be used throughout the question.

Year	1998	1999
Quarter 1	163	165
Quarter 2	265	293
Quarter 3	409	469
Quarter 4	186	196

(a) Plot the data together with a suitable moving average.   (*6 marks*)

(b) Predict the expenditure in Quarter 1 2000. Indicate the method you have used. (*5 marks*)

(c) Comment on your method of forecasting, given that the actual figure for Quarter 1, 2000 was £151 million. (*3 marks*)

5 An airline sells 165 tickets for a flight which can only accommodate 159 passengers. Past experience suggests that the probability of a passenger who has bought a ticket not turning up for the flight may be modelled by a binomial distribution with $p = 0.04$.

(a) Use a suitable approximation to find the probability that all the passengers who turn up can be accommodated. (*7 marks*)

(b) Give a reason why the binomial distribution may not provide a suitable model for the number of passengers who fail to turn up for the flight. (*2 marks*)

(c) State whether a Poisson distribution, a normal distribution or neither would be a suitable approximation for a binomial distribution with the following parameters

　(i) $n = 180$, $p = 0.001$

　(ii) $n = 12$, $p = 0.4$

　(iii) $n = 108$, $p = 0.99$ (*6 marks*)

Appendix

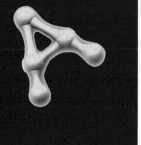

Table 1 Cumulative binomial distribution function

The tabulated value is $P(X \leqslant x)$, where X has a binomial distribution with parameters n and p.

n	x \ p	0.01	0.02	0.03	0.04	0.05	0.06	0.07	0.08	0.09	0.10	0.15	0.20	0.25	0.30	0.35	0.40	0.45	0.50	p \ x
n = 2	0	0.9801	0.9604	0.9409	0.9216	0.9025	0.8836	0.8649	0.8464	0.8281	0.8100	0.7225	0.6400	0.5625	0.4900	0.4225	0.3600	0.3025	0.2500	0
	1	0.9999	0.9996	0.9991	0.9984	0.9975	0.9964	0.9951	0.9936	0.9919	0.9900	0.9775	0.9600	0.9375	0.9100	0.8775	0.8400	0.7975	0.7500	1
	2	1.0000	1.0000	1.0000	1.0000	1.0000	1.0000	1.0000	1.0000	1.0000	1.0000	1.0000	1.0000	1.0000	1.0000	1.0000	1.0000	1.0000	1.0000	2
n = 3	0	0.9703	0.9412	0.9127	0.8847	0.8574	0.8306	0.8044	0.7787	0.7536	0.7290	0.6141	0.5120	0.4219	0.3430	0.2746	0.2160	0.1664	0.1250	0
	1	0.9997	0.9988	0.9974	0.9953	0.9928	0.9896	0.9860	0.9818	0.9772	0.9720	0.9393	0.8960	0.8438	0.7840	0.7183	0.6480	0.5748	0.5000	1
	2	1.0000	1.0000	1.0000	0.9999	0.9999	0.9998	0.9997	0.9995	0.9993	0.9990	0.9966	0.9920	0.9844	0.9730	0.9571	0.9360	0.9089	0.8750	2
	3				1.0000	1.0000	1.0000	1.0000	1.0000	1.0000	1.0000	1.0000	1.0000	1.0000	1.0000	1.0000	1.0000	1.0000	1.0000	3
n = 4	0	0.9606	0.9224	0.8853	0.8493	0.8145	0.7807	0.7481	0.7164	0.6857	0.6561	0.5220	0.4096	0.3164	0.2401	0.1785	0.1296	0.0915	0.0625	0
	1	0.9994	0.9977	0.9948	0.9909	0.9860	0.9801	0.9733	0.9656	0.9570	0.9477	0.8905	0.8192	0.7383	0.6517	0.5630	0.4752	0.3910	0.3125	1
	2	1.0000	1.0000	0.9999	0.9998	0.9995	0.9992	0.9987	0.9981	0.9973	0.9963	0.9880	0.9728	0.9492	0.9163	0.8735	0.8208	0.7585	0.6875	2
	3			1.0000	1.0000	1.0000	1.0000	1.0000	1.0000	0.9999	0.9999	0.9995	0.9984	0.9961	0.9919	0.9850	0.9744	0.9590	0.9375	3
	4									1.0000	1.0000	1.0000	1.0000	1.0000	1.0000	1.0000	1.0000	1.0000	1.0000	4
n = 5	0	0.9510	0.9039	0.8587	0.8154	0.7738	0.7339	0.6957	0.6591	0.6240	0.5905	0.4437	0.3277	0.2373	0.1681	0.1160	0.0778	0.0503	0.0313	0
	1	0.9990	0.9962	0.9915	0.9852	0.9774	0.9681	0.9575	0.9456	0.9326	0.9185	0.8352	0.7373	0.6328	0.5282	0.4284	0.3370	0.2562	0.1875	1
	2	1.0000	0.9999	0.9997	0.9994	0.9988	0.9980	0.9969	0.9955	0.9937	0.9914	0.9734	0.9421	0.8965	0.8369	0.7648	0.6826	0.5931	0.5000	2
	3		1.0000	1.0000	1.0000	1.0000	0.9999	0.9999	0.9998	0.9997	0.9995	0.9978	0.9933	0.9844	0.9692	0.9460	0.9130	0.8688	0.8125	3
	4						1.0000	1.0000	1.0000	1.0000	1.0000	0.9999	0.9997	0.9990	0.9976	0.9947	0.9898	0.9815	0.9688	4
	5											1.0000	1.0000	1.0000	1.0000	1.0000	1.0000	1.0000	1.0000	5
n = 6	0	0.9415	0.8858	0.8330	0.7828	0.7351	0.6899	0.6470	0.6064	0.5679	0.5314	0.3771	0.2621	0.1780	0.1176	0.0754	0.0467	0.0277	0.0156	0
	1	0.9985	0.9943	0.9875	0.9784	0.9672	0.9541	0.9392	0.9227	0.9048	0.8857	0.7765	0.6554	0.5339	0.4202	0.3191	0.2333	0.1636	0.1094	1
	2	1.0000	0.9998	0.9995	0.9988	0.9978	0.9962	0.9942	0.9915	0.9882	0.9842	0.9527	0.9011	0.8306	0.7443	0.6471	0.5443	0.4415	0.3438	2
	3		1.0000	1.0000	1.0000	0.9999	0.9998	0.9997	0.9995	0.9992	0.9987	0.9941	0.9830	0.9624	0.9295	0.8826	0.8208	0.7447	0.6563	3
	4					1.0000	1.0000	1.0000	1.0000	1.0000	0.9999	0.9996	0.9984	0.9954	0.9891	0.9777	0.9590	0.9308	0.8906	4
	5										1.0000	1.0000	0.9999	0.9998	0.9993	0.9982	0.9959	0.9917	0.9844	5
	6												1.0000	1.0000	1.0000	1.0000	1.0000	1.0000	1.0000	6
n = 7	0	0.9321	0.8681	0.8080	0.7514	0.6983	0.6485	0.6017	0.5578	0.5168	0.4783	0.3206	0.2097	0.1335	0.0824	0.0490	0.0280	0.0152	0.0078	0
	1	0.9980	0.9921	0.9829	0.9706	0.9556	0.9382	0.9187	0.8974	0.8745	0.8503	0.7166	0.5767	0.4449	0.3294	0.2338	0.1586	0.1024	0.0625	1
	2	1.0000	0.9997	0.9991	0.9980	0.9962	0.9937	0.9903	0.9860	0.9807	0.9743	0.9262	0.8520	0.7564	0.6471	0.5323	0.4199	0.3164	0.2266	2
	3		1.0000	1.0000	0.9999	0.9998	0.9996	0.9993	0.9988	0.9982	0.9973	0.9879	0.9667	0.9294	0.8740	0.8002	0.7102	0.6083	0.5000	3
	4				1.0000	1.0000	1.0000	1.0000	0.9999	0.9999	0.9998	0.9988	0.9953	0.9871	0.9712	0.9444	0.9037	0.8471	0.7734	4
	5								1.0000	1.0000	1.0000	0.9999	0.9996	0.9987	0.9962	0.9910	0.9812	0.9643	0.9375	5
	6											1.0000	1.0000	0.9999	0.9998	0.9994	0.9984	0.9963	0.9922	6
	7													1.0000	1.0000	1.0000	1.0000	1.0000	1.0000	7
n = 8	0	0.9227	0.8508	0.7837	0.7214	0.6634	0.6096	0.5596	0.5132	0.4703	0.4305	0.2725	0.1678	0.1001	0.0576	0.0319	0.0168	0.0084	0.0039	0
	1	0.9973	0.9897	0.9777	0.9619	0.9428	0.9208	0.8965	0.8702	0.8423	0.8131	0.6572	0.5033	0.3671	0.2553	0.1691	0.1064	0.0632	0.0352	1
	2	0.9999	0.9996	0.9987	0.9969	0.9942	0.9904	0.9853	0.9789	0.9711	0.9619	0.8948	0.7969	0.6785	0.5518	0.4278	0.3154	0.2201	0.1445	2
	3	1.0000	1.0000	0.9999	0.9998	0.9996	0.9993	0.9987	0.9978	0.9966	0.9950	0.9786	0.9437	0.8862	0.8059	0.7064	0.5941	0.4770	0.3633	3
	4			1.0000	1.0000	1.0000	1.0000	0.9999	0.9999	0.9997	0.9996	0.9971	0.9896	0.9727	0.9420	0.8939	0.8263	0.7396	0.6367	4
	5							1.0000	1.0000	1.0000	1.0000	0.9998	0.9988	0.9958	0.9887	0.9747	0.9502	0.9115	0.8555	5
	6											1.0000	0.9999	0.9996	0.9987	0.9964	0.9915	0.9819	0.9648	6
	7												1.0000	1.0000	0.9999	0.9998	0.9993	0.9983	0.9961	7
	8														1.0000	1.0000	1.0000	1.0000	1.0000	8

Table 1 Cumulative binomial distribution function (cont.)

n = 9

x \ p	0.01	0.02	0.03	0.04	0.05	0.06	0.07	0.08	0.09	0.10	0.15	0.20	0.25	0.30	0.35	0.40	0.45	0.50	p \ x
0	0.9135	0.8337	0.7602	0.6925	0.6302	0.5730	0.5204	0.4722	0.4279	0.3874	0.2316	0.1342	0.0751	0.0404	0.0207	0.0101	0.0046	0.0020	0
1	0.9966	0.9869	0.9718	0.9522	0.9288	0.9022	0.8729	0.8417	0.8088	0.7748	0.5995	0.4362	0.3003	0.1960	0.1211	0.0705	0.0385	0.0195	1
2	0.9999	0.9994	0.9980	0.9955	0.9916	0.9862	0.9791	0.9702	0.9595	0.9470	0.8591	0.7382	0.6007	0.4628	0.3373	0.2318	0.1495	0.0898	2
3	1.0000	1.0000	0.9999	0.9997	0.9994	0.9987	0.9977	0.9963	0.9943	0.9917	0.9661	0.9144	0.8343	0.7297	0.6089	0.4826	0.3614	0.2539	3
4			1.0000	1.0000	1.0000	0.9999	0.9998	0.9997	0.9995	0.9991	0.9944	0.9804	0.9511	0.9012	0.8283	0.7334	0.6214	0.5000	4
5						1.0000	1.0000	1.0000	1.0000	0.9999	0.9994	0.9969	0.9900	0.9747	0.9464	0.9006	0.8342	0.7461	5
6										1.0000	1.0000	0.9997	0.9987	0.9957	0.9888	0.9750	0.9502	0.9102	6
7												1.0000	0.9999	0.9996	0.9986	0.9962	0.9909	0.9805	7
8													1.0000	1.0000	0.9999	0.9997	0.9992	0.9980	8
9															1.0000	1.0000	1.0000	1.0000	9

n = 10

x	0.01	0.02	0.03	0.04	0.05	0.06	0.07	0.08	0.09	0.10	0.15	0.20	0.25	0.30	0.35	0.40	0.45	0.50	x
0	0.9044	0.8171	0.7374	0.6648	0.5987	0.5386	0.4840	0.4344	0.3894	0.3487	0.1969	0.1074	0.0563	0.0282	0.0135	0.0060	0.0025	0.0010	0
1	0.9957	0.9838	0.9655	0.9418	0.9139	0.8824	0.8483	0.8121	0.7746	0.7361	0.5443	0.3758	0.2440	0.1493	0.0860	0.0464	0.0233	0.0107	1
2	0.9999	0.9991	0.9972	0.9938	0.9885	0.9812	0.9717	0.9599	0.9460	0.9298	0.8202	0.6778	0.5256	0.3828	0.2616	0.1673	0.0996	0.0547	2
3	1.0000	1.0000	0.9999	0.9996	0.9990	0.9980	0.9964	0.9942	0.9912	0.9872	0.9500	0.8791	0.7759	0.6496	0.5138	0.3823	0.2660	0.1719	3
4			1.0000	1.0000	0.9999	0.9998	0.9997	0.9994	0.9990	0.9984	0.9901	0.9672	0.9219	0.8497	0.7515	0.6331	0.5044	0.3770	4
5					1.0000	1.0000	1.0000	1.0000	0.9999	0.9999	0.9986	0.9936	0.9803	0.9527	0.9051	0.8338	0.7384	0.6230	5
6									1.0000	1.0000	0.9999	0.9991	0.9965	0.9894	0.9740	0.9452	0.8980	0.8281	6
7											1.0000	0.9999	0.9996	0.9984	0.9952	0.9877	0.9726	0.9453	7
8												1.0000	1.0000	0.9999	0.9995	0.9983	0.9955	0.9893	8
9														1.0000	1.0000	0.9999	0.9997	0.9990	9
10																1.0000	1.0000	1.0000	10

n = 11

x	0.01	0.02	0.03	0.04	0.05	0.06	0.07	0.08	0.09	0.10	0.15	0.20	0.25	0.30	0.35	0.40	0.45	0.50	x
0	0.8953	0.8007	0.7153	0.6382	0.5688	0.5063	0.4501	0.3996	0.3544	0.3138	0.1673	0.0859	0.0422	0.0198	0.0088	0.0036	0.0014	0.0005	0
1	0.9948	0.9805	0.9587	0.9308	0.8981	0.8618	0.8228	0.7819	0.7399	0.6974	0.4922	0.3221	0.1971	0.1130	0.0606	0.0302	0.0139	0.0059	1
2	0.9998	0.9988	0.9963	0.9917	0.9848	0.9752	0.9630	0.9481	0.9305	0.9104	0.7788	0.6174	0.4552	0.3127	0.2001	0.1189	0.0652	0.0327	2
3	1.0000	1.0000	0.9998	0.9993	0.9984	0.9970	0.9947	0.9915	0.9871	0.9815	0.9306	0.8389	0.7133	0.5696	0.4256	0.2963	0.1911	0.1133	3
4			1.0000	1.0000	0.9999	0.9997	0.9995	0.9990	0.9983	0.9972	0.9841	0.9496	0.8854	0.7897	0.6683	0.5328	0.3971	0.2744	4
5					1.0000	1.0000	1.0000	0.9999	0.9998	0.9997	0.9973	0.9883	0.9657	0.9218	0.8513	0.7535	0.6331	0.5000	5
6								1.0000	1.0000	1.0000	0.9997	0.9980	0.9924	0.9784	0.9499	0.9006	0.8262	0.7256	6
7											1.0000	0.9998	0.9988	0.9957	0.9878	0.9707	0.9390	0.8867	7
8												1.0000	0.9999	0.9994	0.9980	0.9941	0.9852	0.9673	8
9													1.0000	1.0000	0.9998	0.9993	0.9978	0.9941	9
10															1.0000	1.0000	0.9998	0.9995	10
11																		1.0000	11

n = 12

x	0.01	0.02	0.03	0.04	0.05	0.06	0.07	0.08	0.09	0.10	0.15	0.20	0.25	0.30	0.35	0.40	0.45	0.50	x
0	0.8864	0.7847	0.6938	0.6127	0.5404	0.4759	0.4186	0.3677	0.3225	0.2824	0.1422	0.0687	0.0317	0.0138	0.0057	0.0022	0.0008	0.0002	0
1	0.9938	0.9769	0.9514	0.9191	0.8816	0.8405	0.7967	0.7513	0.7052	0.6590	0.4435	0.2749	0.1584	0.0850	0.0424	0.0196	0.0083	0.0032	1
2	0.9998	0.9985	0.9952	0.9893	0.9804	0.9684	0.9532	0.9348	0.9134	0.8891	0.7358	0.5583	0.3907	0.2528	0.1513	0.0834	0.0421	0.0193	2
3	1.0000	0.9999	0.9997	0.9990	0.9978	0.9957	0.9925	0.9880	0.9820	0.9744	0.9078	0.7946	0.6488	0.4925	0.3467	0.2253	0.1345	0.0730	3
4			1.0000	0.9999	0.9998	0.9996	0.9992	0.9984	0.9973	0.9957	0.9761	0.9274	0.8424	0.7237	0.5833	0.4382	0.3044	0.1938	4
5					1.0000	1.0000	1.0000	0.9998	0.9997	0.9995	0.9954	0.9806	0.9456	0.8822	0.7873	0.6652	0.5269	0.3872	5
6								1.0000	1.0000	0.9999	0.9993	0.9961	0.9857	0.9614	0.9154	0.8418	0.7393	0.6128	6
7										1.0000	0.9999	0.9994	0.9972	0.9905	0.9745	0.9427	0.8883	0.8062	7
8											1.0000	0.9999	0.9996	0.9983	0.9944	0.9847	0.9644	0.9270	8
9												1.0000	1.0000	0.9998	0.9992	0.9972	0.9921	0.9807	9
10														1.0000	0.9999	0.9997	0.9989	0.9968	10
11															1.0000	1.0000	0.9999	0.9998	11
12																		1.0000	12

n = 13

x	0.01	0.02	0.03	0.04	0.05	0.06	0.07	0.08	0.09	0.10	0.15	0.20	0.25	0.30	0.35	0.40	0.45	0.50	x
0	0.8775	0.7690	0.6730	0.5882	0.5133	0.4474	0.3893	0.3383	0.2935	0.2542	0.1209	0.0550	0.0238	0.0097	0.0037	0.0013	0.0004	0.0001	0
1	0.9928	0.9730	0.9436	0.9068	0.8646	0.8186	0.7702	0.7206	0.6707	0.6213	0.3983	0.2336	0.1267	0.0637	0.0296	0.0126	0.0049	0.0017	1
2	0.9997	0.9980	0.9938	0.9865	0.9755	0.9608	0.9422	0.9201	0.8946	0.8661	0.6920	0.5017	0.3326	0.2025	0.1132	0.0579	0.0269	0.0112	2
3	1.0000	0.9999	0.9995	0.9986	0.9969	0.9940	0.9897	0.9837	0.9758	0.9658	0.8820	0.7473	0.5843	0.4206	0.2783	0.1686	0.0929	0.0461	3
4			1.0000	0.9999	0.9997	0.9993	0.9987	0.9976	0.9959	0.9935	0.9658	0.9009	0.7940	0.6543	0.5005	0.3530	0.2279	0.1334	4
5					1.0000	1.0000	1.0000	0.9999	0.9997	0.9991	0.9925	0.9700	0.9198	0.8346	0.7159	0.5744	0.4268	0.2905	5
6									1.0000	0.9999	0.9987	0.9930	0.9757	0.9376	0.8705	0.7712	0.6437	0.5000	6
7										1.0000	0.9998	0.9988	0.9944	0.9818	0.9538	0.9023	0.8212	0.7095	7
8											1.0000	0.9998	0.9990	0.9960	0.9874	0.9679	0.9302	0.8666	8
9												1.0000	0.9999	0.9993	0.9975	0.9922	0.9797	0.9539	9
10													1.0000	0.9999	0.9997	0.9987	0.9959	0.9888	10
11														1.0000	1.0000	0.9999	0.9995	0.9983	11
12																1.0000	1.0000	0.9999	12
13																		1.0000	13

Table I Cumulative binomial distribution function (cont.)

x \ p	0.01	0.02	0.03	0.04	0.05	0.06	0.07	0.08	0.09	0.10	0.15	0.20	0.25	0.30	0.35	0.40	0.45	0.50	p / x
n = 14 0	0.8687	0.7536	0.6528	0.5647	0.4877	0.4205	0.3620	0.3112	0.2670	0.2288	0.1028	0.0440	0.0178	0.0068	0.0024	0.0008	0.0002	0.0001	0
1	0.9916	0.9690	0.9355	0.8941	0.8470	0.7963	0.7436	0.6900	0.6368	0.5846	0.3567	0.1979	0.1010	0.0475	0.0205	0.0081	0.0029	0.0009	1
2	0.9997	0.9975	0.9923	0.9833	0.9699	0.9522	0.9302	0.9042	0.8745	0.8416	0.6479	0.4481	0.2811	0.1608	0.0839	0.0398	0.0170	0.0065	2
3	1.0000	0.9999	0.9994	0.9981	0.9958	0.9920	0.9864	0.9786	0.9685	0.9559	0.8535	0.6982	0.5213	0.3552	0.2205	0.1243	0.0632	0.0287	3
4		1.0000	1.0000	0.9998	0.9996	0.9990	0.9980	0.9965	0.9941	0.9908	0.9533	0.8702	0.7415	0.5842	0.4227	0.2793	0.1672	0.0898	4
5				1.0000	1.0000	0.9999	0.9998	0.9996	0.9992	0.9985	0.9885	0.9561	0.8883	0.7805	0.6405	0.4859	0.3373	0.2120	5
6						1.0000	1.0000	1.0000	0.9999	0.9998	0.9978	0.9884	0.9617	0.9067	0.8164	0.6925	0.5461	0.3953	6
7									1.0000	1.0000	0.9997	0.9976	0.9897	0.9685	0.9247	0.8499	0.7414	0.6047	7
8											1.0000	0.9996	0.9978	0.9917	0.9757	0.9417	0.8811	0.7880	8
9												1.0000	0.9997	0.9983	0.9940	0.9825	0.9574	0.9102	9
10													1.0000	0.9998	0.9989	0.9961	0.9886	0.9713	10
11														1.0000	0.9999	0.9994	0.9978	0.9935	11
12															1.0000	0.9999	0.9997	0.9991	12
13																1.0000	1.0000	0.9999	13
14																		1.0000	14
n = 15 0	0.8601	0.7386	0.6333	0.5421	0.4633	0.3953	0.3367	0.2863	0.2430	0.2059	0.0874	0.0352	0.0134	0.0047	0.0016	0.0005	0.0001	0.0000	0
1	0.9904	0.9647	0.9270	0.8809	0.8290	0.7738	0.7168	0.6597	0.6035	0.5490	0.3186	0.1671	0.0802	0.0353	0.0142	0.0052	0.0017	0.0005	1
2	0.9996	0.9970	0.9906	0.9797	0.9638	0.9429	0.9171	0.8870	0.8531	0.8159	0.6042	0.3980	0.2361	0.1268	0.0617	0.0271	0.0107	0.0037	2
3	1.0000	0.9998	0.9992	0.9976	0.9945	0.9896	0.9825	0.9727	0.9601	0.9444	0.8227	0.6482	0.4613	0.2969	0.1727	0.0905	0.0424	0.0176	3
4		1.0000	0.9999	0.9998	0.9994	0.9986	0.9972	0.9950	0.9918	0.9873	0.9383	0.8358	0.6865	0.5155	0.3519	0.2173	0.1204	0.0592	4
5			1.0000	1.0000	0.9999	0.9999	0.9997	0.9993	0.9987	0.9978	0.9832	0.9389	0.8516	0.7216	0.5643	0.4032	0.2608	0.1509	5
6					1.0000	1.0000	1.0000	0.9999	0.9998	0.9997	0.9964	0.9819	0.9434	0.8689	0.7548	0.6098	0.4522	0.3036	6
7								1.0000	1.0000	1.0000	0.9994	0.9958	0.9827	0.9500	0.8868	0.7869	0.6535	0.5000	7
8											0.9999	0.9992	0.9958	0.9848	0.9578	0.9050	0.8182	0.6964	8
9											1.0000	0.9999	0.9992	0.9963	0.9876	0.9662	0.9231	0.8491	9
10												1.0000	0.9999	0.9993	0.9972	0.9907	0.9745	0.9408	10
11													1.0000	0.9999	0.9995	0.9981	0.9937	0.9824	11
12														1.0000	0.9999	0.9997	0.9989	0.9963	12
13															1.0000	1.0000	0.9999	0.9995	13
14																	1.0000	1.0000	14
n = 20 0	0.8179	0.6676	0.5438	0.4420	0.3585	0.2901	0.2342	0.1887	0.1516	0.1216	0.0388	0.0115	0.0032	0.0008	0.0002	0.0000	0.0000	0.0000	0
1	0.9831	0.9401	0.8802	0.8103	0.7358	0.6605	0.5869	0.5169	0.4516	0.3917	0.1756	0.0692	0.0243	0.0076	0.0021	0.0005	0.0001	0.0000	1
2	0.9990	0.9929	0.9790	0.9561	0.9245	0.8850	0.8390	0.7879	0.7334	0.6769	0.4049	0.2061	0.0913	0.0355	0.0121	0.0036	0.0009	0.0002	2
3	1.0000	0.9994	0.9973	0.9926	0.9841	0.9710	0.9529	0.9294	0.9007	0.8670	0.6477	0.4114	0.2252	0.1071	0.0444	0.0160	0.0049	0.0013	3
4		1.0000	0.9997	0.9990	0.9974	0.9944	0.9893	0.9817	0.9710	0.9568	0.8298	0.6296	0.4148	0.2375	0.1182	0.0510	0.0189	0.0059	4
5			1.0000	0.9999	0.9997	0.9991	0.9981	0.9962	0.9932	0.9887	0.9327	0.8042	0.6172	0.4164	0.2454	0.1256	0.0553	0.0207	5
6				1.0000	1.0000	0.9999	0.9997	0.9994	0.9987	0.9976	0.9781	0.9133	0.7858	0.6080	0.4166	0.2500	0.1299	0.0577	6
7						1.0000	1.0000	0.9999	0.9998	0.9996	0.9941	0.9679	0.8982	0.7723	0.6010	0.4159	0.2520	0.1316	7
8								1.0000	1.0000	0.9999	0.9987	0.9900	0.9591	0.8867	0.7624	0.5956	0.4143	0.2517	8
9										1.0000	0.9998	0.9974	0.9861	0.9520	0.8782	0.7553	0.5914	0.4119	9
10											1.0000	0.9994	0.9961	0.9829	0.9468	0.8725	0.7507	0.5881	10
11												0.9999	0.9991	0.9949	0.9804	0.9435	0.8692	0.7483	11
12												1.0000	0.9998	0.9987	0.9940	0.9790	0.9420	0.8684	12
13													1.0000	0.9997	0.9985	0.9935	0.9786	0.9423	13
14														1.0000	0.9997	0.9984	0.9936	0.9793	14
15															1.0000	0.9997	0.9985	0.9941	15
16																1.0000	0.9997	0.9987	16
17																	1.0000	0.9998	17
18																		1.0000	18
n = 25 0	0.7778	0.6035	0.4670	0.3604	0.2774	0.2129	0.1630	0.1244	0.0946	0.0718	0.0172	0.0038	0.0008	0.0001	0.0000	0.0000	0.0000	0.0000	0
1	0.9742	0.9114	0.8280	0.7358	0.6424	0.5527	0.4696	0.3947	0.3286	0.2712	0.0931	0.0274	0.0070	0.0016	0.0003	0.0001	0.0000	0.0000	1
2	0.9980	0.9868	0.9620	0.9235	0.8729	0.8129	0.7466	0.6768	0.6063	0.5371	0.2537	0.0982	0.0321	0.0090	0.0021	0.0004	0.0001	0.0000	2
3	0.9999	0.9986	0.9938	0.9835	0.9659	0.9402	0.9064	0.8649	0.8169	0.7636	0.4711	0.2340	0.0962	0.0332	0.0097	0.0024	0.0005	0.0001	3
4	1.0000	0.9999	0.9992	0.9972	0.9928	0.9850	0.9726	0.9549	0.9314	0.9020	0.6821	0.4207	0.2137	0.0905	0.0320	0.0095	0.0023	0.0005	4
5		1.0000	0.9999	0.9996	0.9988	0.9969	0.9935	0.9877	0.9790	0.9666	0.8385	0.6167	0.3783	0.1935	0.0826	0.0294	0.0086	0.0020	5
6			1.0000	1.0000	0.9998	0.9995	0.9987	0.9972	0.9946	0.9905	0.9305	0.7800	0.5611	0.3407	0.1734	0.0736	0.0258	0.0073	6
7					1.0000	0.9999	0.9998	0.9995	0.9989	0.9977	0.9745	0.8909	0.7265	0.5118	0.3061	0.1536	0.0639	0.0216	7
8						1.0000	1.0000	0.9999	0.9998	0.9995	0.9920	0.9532	0.8506	0.6769	0.4668	0.2735	0.1340	0.0539	8
9								1.0000	1.0000	0.9999	0.9979	0.9827	0.9287	0.8106	0.6303	0.4246	0.2424	0.1148	9
10										1.0000	0.9995	0.9944	0.9703	0.9022	0.7712	0.5858	0.3843	0.2122	10
11											0.9999	0.9985	0.9893	0.9558	0.8746	0.7323	0.5426	0.3450	11
12											1.0000	0.9996	0.9966	0.9825	0.9396	0.8462	0.6937	0.5000	12
13												0.9999	0.9991	0.9940	0.9745	0.9222	0.8173	0.6550	13
14												1.0000	0.9998	0.9982	0.9907	0.9656	0.9040	0.7878	14
15													1.0000	0.9995	0.9971	0.9868	0.9560	0.8852	15
16														0.9999	0.9992	0.9957	0.9826	0.9461	16
17														1.0000	0.9998	0.9988	0.9942	0.9784	17
18															1.0000	0.9997	0.9984	0.9927	18
19																0.9999	0.9996	0.9980	19
20																1.0000	0.9999	0.9995	20
21																	1.0000	0.9999	21
22																		1.0000	22

Table 1 Cumulative binomial distribution function (cont.)

x	p 0.01	0.02	0.03	0.04	0.05	0.06	0.07	0.08	0.09	0.10	0.15	0.20	0.25	0.30	0.35	0.40	0.45	0.50	p x
n = 30 0	0.7397	0.5455	0.4010	0.2939	0.2146	0.1563	0.1134	0.0820	0.0591	0.0424	0.0076	0.0012	0.0002	0.0000	0.0000	0.0000	0.0000	0.0000	0
1	0.9639	0.8795	0.7731	0.6612	0.5535	0.4555	0.3694	0.2958	0.2343	0.1837	0.0480	0.0105	0.0020	0.0003	0.0000	0.0000	0.0000	0.0000	1
2	0.9967	0.9783	0.9399	0.8831	0.8122	0.7324	0.6487	0.5654	0.4855	0.4114	0.1514	0.0442	0.0106	0.0021	0.0003	0.0000	0.0000	0.0000	2
3	0.9998	0.9971	0.9881	0.9694	0.9392	0.8974	0.8450	0.7842	0.7175	0.6474	0.3217	0.1227	0.0374	0.0093	0.0019	0.0003	0.0000	0.0000	3
4	1.0000	0.9997	0.9982	0.9937	0.9844	0.9685	0.9447	0.9126	0.8723	0.8245	0.5245	0.2552	0.0979	0.0302	0.0075	0.0015	0.0002	0.0000	4
5		1.0000	0.9998	0.9989	0.9967	0.9921	0.9838	0.9707	0.9519	0.9268	0.7106	0.4275	0.2026	0.0766	0.0233	0.0057	0.0011	0.0002	5
6			1.0000	0.9999	0.9994	0.9983	0.9960	0.9918	0.9848	0.9742	0.8474	0.6070	0.3481	0.1595	0.0586	0.0172	0.0040	0.0007	6
7				1.0000	0.9999	0.9997	0.9992	0.9980	0.9959	0.9922	0.9302	0.7608	0.5143	0.2814	0.1238	0.0435	0.0121	0.0026	7
8					1.0000	1.0000	0.9999	0.9996	0.9990	0.9980	0.9722	0.8713	0.6736	0.4315	0.2247	0.0940	0.0312	0.0081	8
9							1.0000	0.9999	0.9998	0.9995	0.9903	0.9389	0.8034	0.5888	0.3575	0.1763	0.0694	0.0214	9
10								1.0000	1.0000	0.9999	0.9971	0.9744	0.8943	0.7304	0.5078	0.2915	0.1350	0.0494	10
11										1.0000	0.9992	0.9905	0.9493	0.8407	0.6548	0.4311	0.2327	0.1002	11
12											0.9998	0.9969	0.9784	0.9155	0.7802	0.5785	0.3592	0.1808	12
13											1.0000	0.9991	0.9918	0.9599	0.8737	0.7145	0.5025	0.2923	13
14												0.9998	0.9973	0.9831	0.9348	0.8246	0.6448	0.4278	14
15												0.9999	0.9992	0.9936	0.9699	0.9029	0.7691	0.5722	15
16												1.0000	0.9998	0.9979	0.9876	0.9519	0.8644	0.7077	16
17													0.9999	0.9994	0.9955	0.9788	0.9286	0.8192	17
18													1.0000	0.9998	0.9986	0.9917	0.9666	0.8998	18
19														1.0000	0.9996	0.9971	0.9862	0.9506	19
20															0.9999	0.9991	0.9950	0.9786	20
21															1.0000	0.9998	0.9984	0.9919	21
22																1.0000	0.9996	0.9974	22
23																	0.9999	0.9993	23
24																	1.0000	0.9998	24
25																		1.0000	25
n = 40 0	0.6690	0.4457	0.2957	0.1954	0.1285	0.0842	0.0549	0.0356	0.0230	0.0148	0.0015	0.0001	0.0000	0.0000	0.0000	0.0000	0.0000	0.0000	0
1	0.9393	0.8095	0.6615	0.5210	0.3991	0.2990	0.2201	0.1594	0.1140	0.0805	0.0121	0.0015	0.0001	0.0000	0.0000	0.0000	0.0000	0.0000	1
2	0.9925	0.9543	0.8822	0.7855	0.6767	0.5665	0.4625	0.3694	0.2894	0.2228	0.0486	0.0079	0.0010	0.0001	0.0000	0.0000	0.0000	0.0000	2
3	0.9993	0.9918	0.9686	0.9252	0.8619	0.7827	0.6937	0.6007	0.5092	0.4231	0.1302	0.0285	0.0047	0.0006	0.0001	0.0000	0.0000	0.0000	3
4	1.0000	0.9988	0.9933	0.9790	0.9520	0.9104	0.8546	0.7868	0.7103	0.6290	0.2633	0.0759	0.0160	0.0026	0.0003	0.0000	0.0000	0.0000	4
5		0.9999	0.9988	0.9951	0.9861	0.9691	0.9419	0.9033	0.8535	0.7937	0.4325	0.1613	0.0433	0.0086	0.0013	0.0001	0.0000	0.0000	5
6		1.0000	0.9998	0.9990	0.9966	0.9909	0.9801	0.9624	0.9361	0.9005	0.6067	0.2859	0.0962	0.0238	0.0044	0.0006	0.0001	0.0000	6
7			1.0000	0.9998	0.9993	0.9977	0.9942	0.9873	0.9758	0.9581	0.7559	0.4371	0.1820	0.0553	0.0124	0.0021	0.0002	0.0000	7
8				1.0000	0.9999	0.9995	0.9985	0.9963	0.9919	0.9845	0.8646	0.5931	0.2998	0.1110	0.0303	0.0061	0.0009	0.0001	8
9					1.0000	0.9999	0.9997	0.9990	0.9976	0.9949	0.9328	0.7318	0.4395	0.1959	0.0644	0.0156	0.0027	0.0003	9
10						1.0000	0.9999	0.9998	0.9994	0.9985	0.9701	0.8392	0.5839	0.3087	0.1215	0.0352	0.0074	0.0011	10
11							1.0000	1.0000	0.9999	0.9996	0.9880	0.9125	0.7151	0.4406	0.2053	0.0709	0.0179	0.0032	11
12									1.0000	0.9999	0.9957	0.9568	0.8209	0.5772	0.3143	0.1285	0.0386	0.0083	12
13										1.0000	0.9986	0.9806	0.8968	0.7032	0.4408	0.2112	0.0751	0.0192	13
14											0.9996	0.9921	0.9456	0.8074	0.5721	0.3174	0.1326	0.0403	14
15											0.9999	0.9971	0.9738	0.8849	0.6946	0.4402	0.2142	0.0769	15
16											1.0000	0.9990	0.9884	0.9367	0.7978	0.5681	0.3185	0.1341	16
17												0.9997	0.9953	0.9680	0.8761	0.6885	0.4391	0.2148	17
18												0.9999	0.9983	0.9852	0.9301	0.7911	0.5651	0.3179	18
19												1.0000	0.9994	0.9937	0.9637	0.8702	0.6844	0.4373	19
20													0.9998	0.9976	0.9827	0.9256	0.7870	0.5627	20
21													1.0000	0.9991	0.9925	0.9608	0.8669	0.6821	21
22														0.9997	0.9970	0.9811	0.9233	0.7852	22
23														0.9999	0.9989	0.9917	0.9595	0.8659	23
24														1.0000	0.9996	0.9966	0.9804	0.9231	24
25															0.9999	0.9988	0.9914	0.9597	25
26															1.0000	0.9996	0.9966	0.9808	26
27																0.9999	0.9988	0.9917	27
28																1.0000	0.9996	0.9968	28
29																	0.9999	0.9989	29
30																	1.0000	0.9997	30
31																		0.9999	31
32																		1.0000	32

Table 1 Cumulative binomial distribution function (cont.)

x	p=0.01	0.02	0.03	0.04	0.05	0.06	0.07	0.08	0.09	0.10	0.15	0.20	0.25	0.30	0.35	0.40	0.45	0.50	x
n=50 0	0.6050	0.3642	0.2181	0.1299	0.0769	0.0453	0.0266	0.0155	0.0090	0.0052	0.0003	0.0000	0.0000	0.0000	0.0000	0.0000	0.0000	0.0000	0
1	0.9106	0.7358	0.5553	0.4005	0.2794	0.1900	0.1265	0.0827	0.0532	0.0338	0.0029	0.0002	0.0000	0.0000	0.0000	0.0000	0.0000	0.0000	1
2	0.9862	0.9216	0.8108	0.6767	0.5405	0.4162	0.3108	0.2260	0.1605	0.1117	0.0142	0.0013	0.0001	0.0000	0.0000	0.0000	0.0000	0.0000	2
3	0.9984	0.9822	0.9372	0.8609	0.7604	0.6473	0.5327	0.4253	0.3303	0.2503	0.0460	0.0057	0.0005	0.0000	0.0000	0.0000	0.0000	0.0000	3
4	0.9999	0.9968	0.9832	0.9510	0.8964	0.8206	0.7290	0.6290	0.5277	0.4312	0.1121	0.0185	0.0021	0.0002	0.0000	0.0000	0.0000	0.0000	4
5	1.0000	0.9995	0.9963	0.9856	0.9622	0.9224	0.8650	0.7919	0.7072	0.6161	0.2194	0.0480	0.0070	0.0007	0.0001	0.0000	0.0000	0.0000	5
6		0.9999	0.9993	0.9964	0.9882	0.9711	0.9417	0.8981	0.8404	0.7702	0.3613	0.1034	0.0194	0.0025	0.0002	0.0000	0.0000	0.0000	6
7		1.0000	0.9999	0.9992	0.9968	0.9906	0.9780	0.9562	0.9232	0.8779	0.5188	0.1904	0.0453	0.0073	0.0008	0.0001	0.0000	0.0000	7
8			1.0000	0.9999	0.9992	0.9973	0.9927	0.9833	0.9672	0.9421	0.6681	0.3073	0.0916	0.0183	0.0025	0.0002	0.0000	0.0000	8
9				1.0000	0.9998	0.9993	0.9978	0.9944	0.9875	0.9755	0.7911	0.4437	0.1637	0.0402	0.0067	0.0008	0.0001	0.0000	9
10					1.0000	0.9998	0.9994	0.9983	0.9957	0.9906	0.8801	0.5836	0.2622	0.0789	0.0160	0.0022	0.0002	0.0000	10
11						1.0000	0.9999	0.9995	0.9987	0.9968	0.9372	0.7107	0.3816	0.1390	0.0342	0.0057	0.0006	0.0000	11
12							1.0000	0.9999	0.9996	0.9990	0.9699	0.8139	0.5110	0.2229	0.0661	0.0133	0.0018	0.0002	12
13								1.0000	0.9999	0.9997	0.9868	0.8894	0.6370	0.3279	0.1163	0.0280	0.0045	0.0005	13
14									1.0000	0.9999	0.9947	0.9393	0.7481	0.4468	0.1878	0.0540	0.0104	0.0013	14
15										1.0000	0.9981	0.9692	0.8369	0.5692	0.2801	0.0955	0.0220	0.0033	15
16											0.9993	0.9856	0.9017	0.6839	0.3889	0.1561	0.0427	0.0077	16
17											0.9998	0.9937	0.9449	0.7822	0.5060	0.2369	0.0765	0.0164	17
18											0.9999	0.9975	0.9713	0.8594	0.6216	0.3356	0.1273	0.0325	18
19											1.0000	0.9991	0.9861	0.9152	0.7264	0.4465	0.1974	0.0595	19
20												0.9997	0.9937	0.9522	0.8139	0.5610	0.2862	0.1013	20
21												0.9999	0.9974	0.9749	0.8813	0.6701	0.3900	0.1611	21
22												1.0000	0.9990	0.9877	0.9290	0.7660	0.5019	0.2399	22
23													0.9996	0.9944	0.9604	0.8438	0.6134	0.3359	23
24													0.9999	0.9976	0.9793	0.9022	0.7160	0.4439	24
25													1.0000	0.9991	0.9900	0.9427	0.8034	0.5561	25
26														0.9997	0.9955	0.9686	0.8721	0.6641	26
27														0.9999	0.9981	0.9840	0.9220	0.7601	27
28														1.0000	0.9993	0.9924	0.9556	0.8389	28
29															0.9997	0.9966	0.9765	0.8987	29
30															0.9999	0.9986	0.9884	0.9405	30
31															1.0000	0.9995	0.9947	0.9675	31
32																0.9998	0.9978	0.9836	32
33																0.9999	0.9991	0.9923	33
34																1.0000	0.9997	0.9967	34
35																	0.9999	0.9987	35
36																	1.0000	0.9995	36
37																		0.9998	37
38																		1.0000	38

Table 2 Cumulative Poisson distribution function

The tabulated value is $P(X \leqslant x)$, where X has a Poisson distribution with mean λ.

x \ λ	0.1	0.2	0.3	0.4	0.5	0.6	0.7	0.8	0.9	1.0	1.2	1.4	1.6	1.8	x
0	0.9048	0.8187	0.7408	0.6703	0.6065	0.5488	0.4966	0.4493	0.4066	0.3679	0.3012	0.2466	0.2019	0.1653	0
1	0.9953	0.9825	0.9631	0.9384	0.9098	0.8781	0.8442	0.8088	0.7725	0.7358	0.6626	0.5918	0.5249	0.4628	1
2	0.9998	0.9989	0.9964	0.9921	0.9856	0.9769	0.9659	0.9526	0.9371	0.9197	0.8795	0.8335	0.7834	0.7306	2
3	1.0000	0.9999	0.9997	0.9992	0.9982	0.9966	0.9942	0.9909	0.9865	0.9810	0.9662	0.9463	0.9212	0.8913	3
4		1.0000	1.0000	0.9999	0.9998	0.9996	0.9992	0.9986	0.9977	0.9963	0.9923	0.9857	0.9763	0.9636	4
5				1.0000	1.0000	1.0000	0.9999	0.9998	0.9997	0.9994	0.9985	0.9968	0.9940	0.9896	5
6							1.0000	1.0000	1.0000	0.9999	0.9997	0.9994	0.9987	0.9974	6
7										1.0000	1.0000	0.9999	0.9997	0.9994	7
8												1.0000	1.0000	0.9999	8
9														1.0000	9

x \ λ	2.0	2.2	2.4	2.6	2.8	3.0	3.2	3.4	3.6	3.8	4.0	4.5	5.0	5.5	x
0	0.1353	0.1108	0.0907	0.0743	0.0608	0.0498	0.0408	0.0334	0.0273	0.0224	0.0183	0.0111	0.0067	0.0041	0
1	0.4060	0.3546	0.3084	0.2674	0.2311	0.1991	0.1712	0.1468	0.1257	0.1074	0.0916	0.0611	0.0404	0.0266	1
2	0.6767	0.6227	0.5697	0.5184	0.4695	0.4232	0.3799	0.3397	0.3027	0.2689	0.2381	0.1736	0.1247	0.0884	2
3	0.8571	0.8194	0.7787	0.7360	0.6919	0.6472	0.6025	0.5584	0.5152	0.4735	0.4335	0.3423	0.2650	0.2017	3
4	0.9473	0.9275	0.9041	0.8774	0.8477	0.8153	0.7806	0.7442	0.7064	0.6678	0.6288	0.5321	0.4405	0.3575	4
5	0.9834	0.9751	0.9643	0.9510	0.9349	0.9161	0.8946	0.8705	0.8441	0.8156	0.7851	0.7029	0.6160	0.5289	5
6	0.9955	0.9925	0.9884	0.9828	0.9756	0.9665	0.9554	0.9421	0.9267	0.9091	0.8893	0.8311	0.7622	0.6860	6
7	0.9989	0.9980	0.9967	0.9947	0.9919	0.9881	0.9832	0.9769	0.9692	0.9599	0.9489	0.9134	0.8666	0.8095	7
8	0.9998	0.9995	0.9991	0.9985	0.9976	0.9962	0.9943	0.9917	0.9883	0.9840	0.9786	0.9597	0.9319	0.8944	8
9	1.0000	0.9999	0.9998	0.9996	0.9993	0.9989	0.9982	0.9973	0.9960	0.9942	0.9919	0.9829	0.9682	0.9462	9
10		1.0000	1.0000	0.9999	0.9998	0.9997	0.9995	0.9992	0.9987	0.9981	0.9972	0.9933	0.9863	0.9747	10
11				1.0000	1.0000	0.9999	0.9999	0.9998	0.9996	0.9994	0.9991	0.9976	0.9945	0.9890	11
12						1.0000	1.0000	0.9999	0.9999	0.9998	0.9997	0.9992	0.9980	0.9955	12
13								1.0000	1.0000	1.0000	0.9999	0.9997	0.9993	0.9983	13
14											1.0000	0.9999	0.9998	0.9994	14
15												1.0000	0.9999	0.9998	15
16													1.0000	0.9999	16
17														1.0000	17

Table 2 Cumulative Poisson distribution function (cont.)

x \ λ	6.0	6.5	7.0	7.5	8.0	8.5	9.0	9.5	10.0	11.0	12.0	13.0	14.0	15.0	λ \ r
0	0.0025	0.0015	0.0009	0.0006	0.0003	0.0002	0.0001	0.0001	0.0000	0.0000	0.0000	0.0000	0.0000	0.0000	0
1	0.0174	0.0113	0.0073	0.0047	0.0030	0.0019	0.0012	0.0008	0.0005	0.0002	0.0001	0.0000	0.0000	0.0000	1
2	0.0620	0.0430	0.0296	0.0203	0.0138	0.0093	0.0062	0.0042	0.0028	0.0012	0.0005	0.0002	0.0001	0.0000	2
3	0.1512	0.1118	0.0818	0.0591	0.0424	0.0301	0.0212	0.0149	0.0103	0.0049	0.0023	0.0011	0.0005	0.0002	3
4	0.2851	0.2237	0.1730	0.1321	0.0996	0.0744	0.0550	0.0403	0.0293	0.0151	0.0076	0.0037	0.0018	0.0009	4
5	0.4457	0.3690	0.3007	0.2414	0.1912	0.1496	0.1157	0.0885	0.0671	0.0375	0.0203	0.0107	0.0055	0.0028	5
6	0.6063	0.5265	0.4497	0.3782	0.3134	0.2562	0.2068	0.1649	0.1301	0.0786	0.0458	0.0259	0.0142	0.0076	6
7	0.7440	0.6728	0.5987	0.5246	0.4530	0.3856	0.3239	0.2687	0.2202	0.1432	0.0895	0.0540	0.0316	0.0180	7
8	0.8472	0.7916	0.7291	0.6620	0.5925	0.5231	0.4557	0.3918	0.3328	0.2320	0.1550	0.0998	0.0621	0.0374	8
9	0.9161	0.8774	0.8305	0.7764	0.7166	0.6530	0.5874	0.5218	0.4579	0.3405	0.2424	0.1658	0.1094	0.0699	9
10	0.9574	0.9332	0.9015	0.8622	0.8159	0.7634	0.7060	0.6453	0.5830	0.4599	0.3472	0.2517	0.1757	0.1185	10
11	0.9799	0.9661	0.9467	0.9208	0.8881	0.8487	0.8030	0.7520	0.6968	0.5793	0.4616	0.3532	0.2600	0.1848	11
12	0.9912	0.9840	0.9730	0.9573	0.9362	0.9091	0.8758	0.8364	0.7916	0.6887	0.5760	0.4631	0.3585	0.2676	12
13	0.9964	0.9929	0.9872	0.9784	0.9658	0.9486	0.9261	0.8981	0.8645	0.7813	0.6815	0.5730	0.4644	0.3632	13
14	0.9986	0.9970	0.9943	0.9897	0.9827	0.9726	0.9585	0.9400	0.9165	0.8540	0.7720	0.6751	0.5704	0.4657	14
15	0.9995	0.9988	0.9976	0.9954	0.9918	0.9862	0.9780	0.9665	0.9513	0.9074	0.8444	0.7636	0.6694	0.5681	15
16	0.9998	0.9996	0.9990	0.9980	0.9963	0.9934	0.9889	0.9823	0.9730	0.9441	0.8987	0.8355	0.7559	0.6641	16
17	0.9999	0.9998	0.9996	0.9992	0.9984	0.9970	0.9947	0.9911	0.9857	0.9678	0.9370	0.8905	0.8272	0.7489	17
18	1.0000	0.9999	0.9999	0.9997	0.9993	0.9987	0.9976	0.9957	0.9928	0.9823	0.9626	0.9302	0.8826	0.8195	18
19		1.0000	1.0000	0.9999	0.9997	0.9995	0.9989	0.9980	0.9965	0.9907	0.9787	0.9573	0.9235	0.8752	19
20				1.0000	0.9999	0.9998	0.9996	0.9991	0.9984	0.9953	0.9884	0.9750	0.9521	0.9170	20
21					1.0000	0.9999	0.9998	0.9996	0.9993	0.9977	0.9939	0.9859	0.9712	0.9469	21
22						1.0000	0.9999	0.9999	0.9997	0.9990	0.9970	0.9924	0.9833	0.9673	22
23							1.0000	0.9999	0.9999	0.9995	0.9985	0.9960	0.9907	0.9805	23
24								1.0000	1.0000	0.9998	0.9993	0.9980	0.9950	0.9888	24
25										0.9999	0.9997	0.9990	0.9974	0.9938	25
26										1.0000	0.9999	0.9995	0.9987	0.9967	26
27											0.9999	0.9998	0.9994	0.9983	27
28											1.0000	0.9999	0.9997	0.9991	28
29												1.0000	0.9999	0.9996	29
30													0.9999	0.9998	30
31													1.0000	0.9999	31
32														1.0000	32

Table 3 Normal distribution function

The table gives the probability p that a normally distributed
random variable Z, with mean = 0 and variance = 1, is less
than or equal to z.

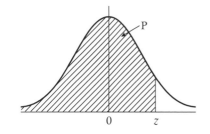

z	0.00	0.01	0.02	0.03	0.04	0.05	0.06	0.07	0.08	0.09	z
0.1	0.50000	0.50399	0.50798	0.51197	0.51595	0.51994	0.52392	0.52790	0.53188	0.53586	0.0
0.1	0.53983	0.54380	0.54776	0.55172	0.55567	0.55962	0.56356	0.56749	0.57142	0.57535	0.1
0.2	0.57926	0.58317	0.58706	0.59095	0.59483	0.59871	0.60257	0.60642	0.61026	0.61409	0.2
0.3	0.61791	0.62172	0.62552	0.62930	0.63307	0.63683	0.64058	0.64431	0.64803	0.65173	0.3
0.4	0.65542	0.65910	0.66276	0.66640	0.67003	0.67364	0.67724	0.68082	0.68439	0.68793	0.4
0.5	0.69146	0.69497	0.69847	0.70194	0.70540	0.70884	0.71226	0.71566	0.71904	0.72240	0.5
0.6	0.72575	0.72907	0.73237	0.73565	0.73891	0.74215	0.74537	0.74857	0.75175	0.75490	0.6
0.7	0.75804	0.76115	0.76424	0.76730	0.77035	0.77337	0.77637	0.77935	0.78230	0.78524	0.7
0.8	0.78814	0.79103	0.79389	0.79673	0.79955	0.80234	0.80511	0.80785	0.81057	0.81327	0.8
0.9	0.81594	0.81859	0.82121	0.82381	0.82639	0.82894	0.83147	0.83398	0.83646	0.83891	0.9
1.0	0.84134	0.84375	0.84614	0.84849	0.85083	0.85314	0.85543	0.85769	0.85993	0.86214	1.0
1.1	0.86433	0.86650	0.86864	0.87076	0.87286	0.87493	0.87698	0.87900	0.88100	0.88298	1.1
1.2	0.88493	0.88686	0.88877	0.89065	0.89251	0.89435	0.89617	0.89796	0.89973	0.90147	1.2
1.3	0.90320	0.90490	0.90658	0.90824	0.90988	0.91149	0.91309	0.91466	0.91621	0.91774	1.3
1.4	0.91924	0.92073	0.92220	0.92364	0.92507	0.92647	0.92785	0.92922	0.93056	0.93189	1.4
1.5	0.93319	0.93448	0.93574	0.93699	0.93822	0.93943	0.94062	0.94179	0.94295	0.94408	1.5
1.6	0.94520	0.94630	0.94738	0.94845	0.94950	0.95053	0.95154	0.95254	0.95352	0.95449	1.6
1.7	0.95543	0.95637	0.95728	0.95818	0.95907	0.95994	0.96080	0.96164	0.96246	0.96327	1.7
1.8	0.96407	0.96485	0.96562	0.96638	0.96712	0.96784	0.96856	0.96926	0.96995	0.97062	1.8
1.9	0.97128	0.97193	0.97257	0.97320	0.97381	0.97441	0.97500	0.97558	0.97615	0.97670	1.9
2.0	0.97725	0.97778	0.97831	0.97882	0.97932	0.97982	0.98030	0.98077	0.98124	0.98169	2.0
2.1	0.98214	0.98257	0.98300	0.98341	0.98382	0.98422	0.98461	0.98500	0.98537	0.98574	2.1
2.2	0.98610	0.98645	0.98679	0.98713	0.98745	0.98778	0.98809	0.98840	0.98870	0.98899	2.2
2.3	0.98928	0.98956	0.98983	0.99010	0.99036	0.99061	0.99086	0.99111	0.99134	0.99158	2.3
2.4	0.99180	0.99202	0.99224	0.99245	0.99266	0.99286	0.99305	0.99324	0.99343	0.99361	2.4
2.5	0.99379	0.99396	0.99413	0.99430	0.99446	0.99461	0.99477	0.99492	0.99506	0.99520	2.5
2.6	0.99534	0.99547	0.99560	0.99573	0.99585	0.99598	0.99609	0.99621	0.99632	0.99643	2.6
2.7	0.99653	0.99664	0.99674	0.99683	0.99693	0.99702	0.99711	0.99720	0.99728	0.99736	2.7
2.8	0.99744	0.99752	0.99760	0.99767	0.99774	0.99781	0.99788	0.99795	0.99801	0.99807	2.8
2.9	0.99813	0.99819	0.99825	0.99831	0.99836	0.99841	0.99846	0.99851	0.99856	0.99861	2.9
3.0	0.99865	0.99869	0.99874	0.99878	0.99882	0.99886	0.99889	0.99893	0.99896	0.99900	3.0
3.1	0.99903	0.99906	0.99910	0.99913	0.99916	0.99918	0.99921	0.99924	0.99926	0.99929	3.1
3.2	0.99931	0.99934	0.99936	0.99938	0.99940	0.99942	0.99944	0.99946	0.99948	0.99950	3.2
3.3	0.99952	0.99953	0.99955	0.99957	0.99958	0.99960	0.99961	0.99962	0.99964	0.99965	3.3
3.4	0.99966	0.99968	0.99969	0.99970	0.99971	0.99972	0.99973	0.99974	0.99975	0.99976	3.4
3.5	0.99977	0.99978	0.99978	0.99979	0.99980	0.99981	0.99981	0.99982	0.99983	0.99983	3.5
3.6	0.99984	0.99985	0.99985	0.99986	0.99986	0.99987	0.99987	0.99988	0.99988	0.99989	3.6
3.7	0.99989	0.99990	0.99990	0.99990	0.99991	0.99991	0.99992	0.99992	0.99992	0.99992	3.7
3.8	0.99993	0.99993	0.99993	0.99994	0.99994	0.99994	0.99994	0.99995	0.99995	0.99995	3.8
3.9	0.99995	0.99995	0.99996	0.99996	0.99996	0.99996	0.99996	0.99996	0.99997	0.99997	3.9

Table 4 Percentage points of the normal distribution

The table gives the values of z satisfying $P(Z \leq z) = p$, where Z is the normally distributed random variable with mean = 0 and variance = 1.

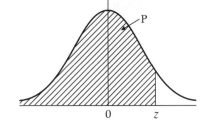

p	0.00	0.01	0.02	0.03	0.04	0.05	0.06	0.07	0.08	0.09	p
0.5	0.0000	0.0251	0.0502	0.0753	0.1004	0.1257	0.1510	0.1764	0.2019	0.2275	**0.5**
0.6	0.2533	0.2793	0.3055	0.3319	0.3585	0.3853	0.4125	0.4399	0.4677	0.4958	**0.6**
0.7	0.5244	0.5534	0.5828	0.6128	0.6433	0.6745	0.7063	0.7388	0.7722	0.8064	**0.7**
0.8	0.8416	0.8779	0.9154	0.9542	0.9945	1.0364	1.0803	1.1264	1.1750	1.2265	**0.8**
0.9	1.2816	1.3408	1.4051	1.4758	1.5548	1.6449	1.7507	1.8808	2.0537	2.3263	**0.9**

p	0.000	0.001	0.002	0.003	0.004	0.005	0.006	0.007	0.008	0.009	p
0.95	1.6449	1.6546	1.6646	1.6747	1.6849	1.6954	1.7060	1.7169	1.7279	1.7392	**0.95**
0.96	1.7507	1.7624	1.7744	1.7866	1.7991	1.8119	1.8250	1.8384	1.8522	1.8663	**0.96**
0.97	1.8808	1.8957	1.9110	1.9268	1.9431	1.9600	1.9774	1.9954	2.0141	2.0335	**0.97**
0.98	2.0537	2.0749	2.0969	2.1201	2.1444	2.1701	2.1973	2.2262	2.2571	2.2904	**0.98**
0.99	2.3263	2.3656	2.4089	2.4573	2.5121	2.5758	2.6521	2.7478	2.8782	3.0902	**0.99**

Answers

EXERCISE 1A

1 (a) (i)

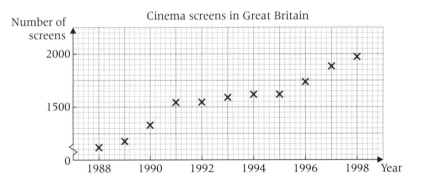

Cinema screens in Great Britain

The trend is upward and approximately linear.

(ii)

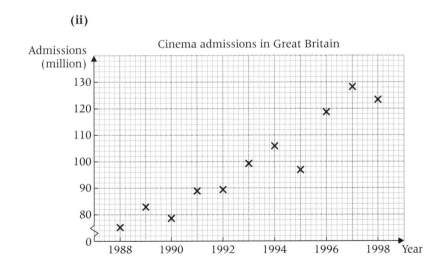

Cinema admissions in Great Britain

The trend is upward and approximately linear.

(b) Both trends are upward and approximately linear but, compared to the number of screens, the admissions show more variability about the trend.

2

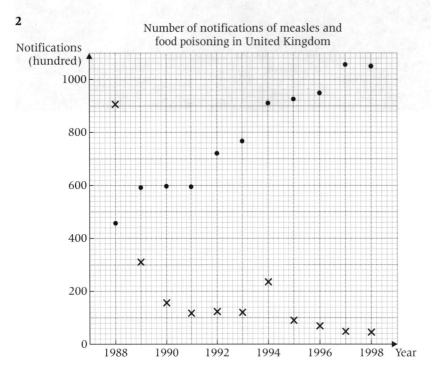

Number of notifications of measles and food poisoning in United Kingdom

Note: as notifications of measles and of food poisoning were both in hundreds it was possible to show them both on the same graph. This is not essential but may be helpful when making comparisons.

(a) Notifications of measles showed a downward non-linear trend. The drop was particularly sharp up to 1990. There were approximately double the number of cases in 1994 than would have been expected from the trend.

(b) Notifications of food poisoning shows an upward, approximately linear trend.

3 (a)

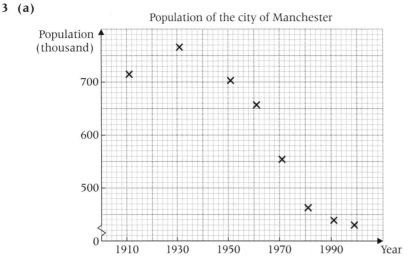

Population of the city of Manchester

Note. Unusually for a time series the points are not evenly spaced along the horizontal axis.

From 1930 the trend is downward but not linear.

(b)

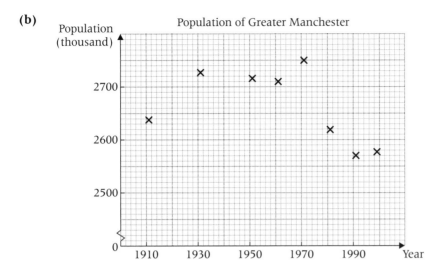

Population of Greater Manchester

Note: it would have been possible to put both series on the same graph. This would have had advantages for making comparisons but it would have been difficult to see the trend in the City population as this is much smaller than the population of Greater Manchester.

There has been some reduction in the population since 1930 but there is no clear trend in this data.

The reduction in the population of the City has been much greater than the reduction in the population of Greater Manchester. The proportion of the population of greater Manchester living in the City has reduced from more than a quarter in 1911 to about a sixth in 1998.

EXERCISE 1B

1

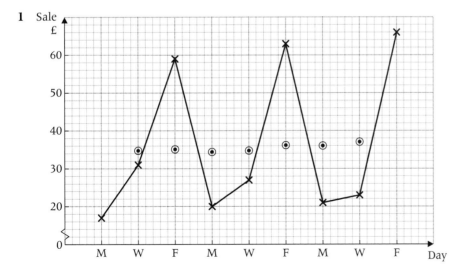

2 (a)

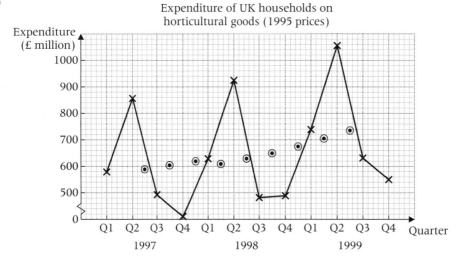

(b) Wednesday, week 2. Attendance was substantially further below the moving average than the other two Wednesdays.

(c) Approximately 290.

(d) The moving average shows an upward approximately linear trend. This cannot continue in the long term since the theatre holds a maximum of 600 people.

3 (a) 152.25 (£152 250); **(b)** 176.3 (£176 300).

4 (a)

Expenditure of UK households on horticultural goods (1995 prices)

(b) Upward, approximately linear.

(c) About £810 000 000 at constant 1995 prices.

(d) Forecast is reasonably close but overestimates the actual expenditure. Not too much should be read into one quarter's figures but the upward trend may be levelling out.

(e) Not much gardening is done in the winter. Highest expenditure is in Quarter 2 when people are preparing to resume.

5 (a)

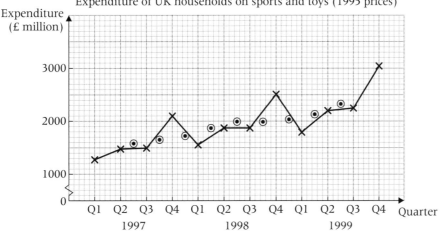

Expenditure of UK households on sports and toys (1995 prices)

(b) Upward, approximately linear.

(c) Predicted moving averages for 2000 in £million, 1995 prices:

Q1 – 2510 Q2 – 2600 Q3 – 2690 Q4 – 2780.

(Note: if $t = 1$ for Q1, 1997, then first moving average corresponds to $t = 2.5$.)

(d) Predicted expenditure 2000 in £million, 1995 prices, estimated to nearest 50:

Q1 – 2250 Q2 – 2550 Q3 – 2600 Q4 – 3250.

(e) Forecast is reasonably close but overestimates the actual expenditure. Not too much should be read into a single quarter but the rate of increase may be reducing.

(f) Highest expenditure is in Quarter 4 which includes Christmas.

EXERCISE IC

1 Possible answers

(a)

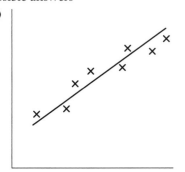

(b)

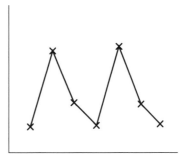

(c)

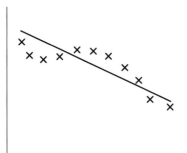

(d)

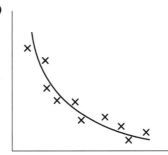

(e)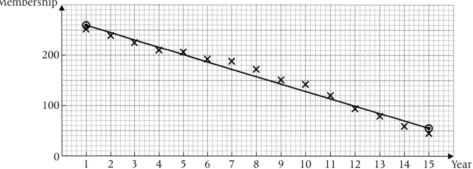

(f)

2 (a) A, **(b)** E, **(c)** F, **(d)** B.

3 (a)

Lodge membership

(b) $y = 276 - 14.8t$.
(c) Short-term variability about a downward, linear trend.
(d) 40.
(e) Graph suggests the actual the value will be below the regression line, say 30.
(f) The regression line would predict a negative membership for 2005 which is impossible.

MIXED EXERCISE

1 (a)

Value of goods exported from UK to Norway

(b) Upward, approximately linear, trend.
(c) Trend would still be upward but the rate of increase would be less.

2 (a)

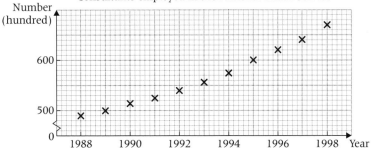

Consultants employed in the National Heath Service

Number of consultants shows an approximately linear upward trend.

(ii)

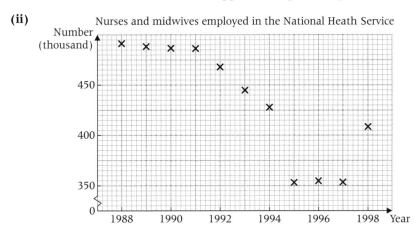

Nurses and midwives employed in the National Heath Service

Although the number of nurses and midwives has reduced there is no consistent trend to the data.

In 1988 there were about ten times as many nurses and midwives employed in the National Health Service as there were consultants. The number of consultants has shown a steady upward trend while the number of nurses and midwives has decreased. In 1998 there were only about six times as many nurses and midwives as consultants.

3 (a)

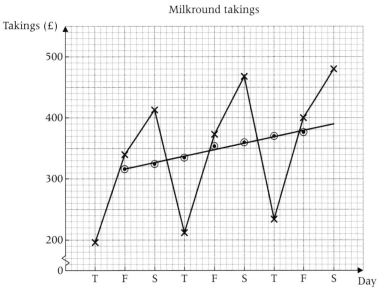

Milkround takings

(b) Predicted moving average for next week:
Tuesday £395, Friday £405, Saturday £415.

(c) Predicted takings for next week:
Tuesday £270, Friday £430, Saturday £520.

(d) It would be foolish to use this small amount of data to predict takings
a year ahead. The trend is unlikely to continue unchanged and even if
it did this amount of data will give only a rough estimate of the trend.
Projecting this estimate a year ahead could lead to major errors.

4 (a)

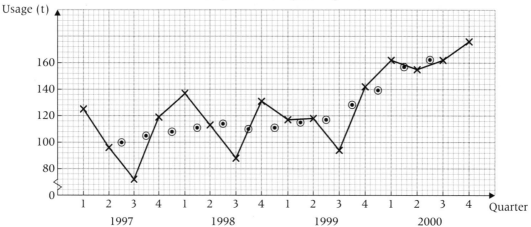

Oil usage of small engineering firm

(b) Quarter 1, 1999, usage is close to moving average. For all other first
quarters the usage is well above the moving average.

(c) Quarter 3, 2000, usage is close to moving average. For other third
quarters the usage is well below the moving average.

5 (a)

Expenditure of UK households on
stationery (1995 prices)

(b) Predicted moving average, £million at 1995 prices, for 2000:

Quarter 1 875.5, Quarter 2 876.0, Quarter 3 876.4, Quarter 4 876.9.

Note: if $t = 1$ for quarter 1 1997, then $t = 2.5$ for first moving
average point.

(c) Predicted actual expenditure, £million at 1995 prices, for 2000

Quarter 1 840, Quarter 2 780, Quarter 3 820, Quarter 4 1060.

(d) Prediction was a slight underestimate but fairly accurate. Method of forecasting seems satisfactory in this case.

6 (a)

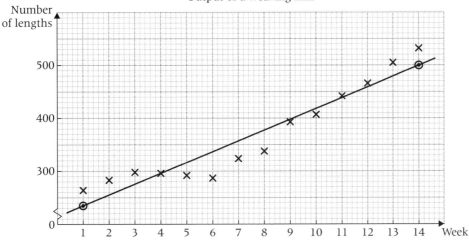

Output of a weaving mill

(b) $y = 212 + 20.5t$.
(c) Short-term variability about an upward linear trend.
(d) 520.
(e) 550.

2 Sampling

EXERCISE 2A

1 Prior knowledge is used to divide the population into strata. Samples are taken from each of the strata in proportion to the size of each of the strata.

Advantage – ensures sample contains same proportion of each of the strata as the population (unlike a random sample).

Disadvantage – more complicated to obtain (or needs relevant prior knowledge).

2 Number males 00 to 11; choose two-digit random numbers; ignore repeats and >11; continue until four obtained; select corresponding males. Choose six of the 18 females in the same way.

3 (a) (i) Stratified sampling will ensure that the sample contains men and women in the same proportion as the population. If there is a difference in attitudes between genders this will ensure the population is fairly represented.
(ii) The additional information needed and extra difficulty of obtaining a stratified sample will be to no purpose if there is no difference in attitude between genders.
(b) age, parent/not parent, employed/unemployed, etc.

EXERCISE 2B

1 Prior knowledge of the population is used to divide it into strata. Samples from each of the strata (usually in proportion to the size of the strata) are taken by any convenient method. Use when relevant prior knowledge is available and random sampling is too difficult or too expensive.

2 Please distribute the questionnaire to nine sixth-form girls, seven sixth-form boys, 19 girls who are not in the sixth form and 20 boys who are not in the sixth form. Provided you stick to these numbers you may ask the first pupils you see or use any other convenient method to select the sample.

3 **(a) (i)** C   **(ii)** A random, B stratified.

   (b) A equally likely, B equally likely, C not equally likely (or depends on the 'convenient method' chosen).

   (c) (i) no possibility of bias,

       (ii) easier to carry out (also ensures all factories fairly represented).

   (d) Both avoid the possibility of bias, but B ensures all factories are fairly represented.

EXERCISE 2C

1 Population is divided into clusters (usually geographically). A sample of clusters is selected at random and then all, or a random sample of, individuals within the chosen clusters form the sample. For example to choose a sample of oil tanker drivers in UK, choose a random sample of oil distribution depots and then a random sample of drivers from the chosen depots. This sample will be more geographically localised than a random sample. It is likely to contain drivers with less varied views (more homogeneous) than the population of all UK tanker drivers.

2 As Question **1**. Use where the population can be divided naturally into clusters and travelling or other difficulties make random or stratified sampling impractical.

   Cluster sampling eliminates the possibility of bias by using random selection. Unlike quota sampling it does not ensure that different strata in the population are fairly represented in the sample.

3 **(a)** Cluster.

   (b) No, all branches would be equally likely to be selected and so members in the smaller branches would have more chance of being included.

   (c) All the branches had an equal number of members.

EXERCISE 2D

1 The sample is taken at regular intervals, for example a local radio reporter interviews every 200th person in a queue waiting to buy tickets for a football match.

2 Number employees 0000 to 2699. Choose a starting point between 0000 and 0053. Select this employee and every 54th thereafter. For example if 0023 is chosen as the starting point choose employees numbered 0023, 0077, 0131, ..., 2669.

3 **(a)** Systematic.

   (b) Yes, all have a probability of $\frac{1}{5}$ of being selected.

   (c) Not all subsets of size 40 could be chosen. For example two friends entering the library together could not both be included.

   (d) Likely to provide a useful estimate. Only unsatisfactory if the average number of books borrowed by customers who come early in the day is different from the average for the population as a whole or if the particular morning chosen is for some reason untypical.

EXERCISE 2E

1 **(a)** Number part time students 00 to 11, select two-digit random numbers; ignore repeats and >11; continue until four numbers chosen; select corresponding students. Choose eight of the full-time students in the same way.

(b) $\bar{x}$, there is a clearly a difference between the average age of part-time and full-time students. The stratified sample will ensure that the two strata are fairly represented in the sample. This might or might not be the case for the random sample.

2 **(a)** Yes.

(b) No, not all subsets of size 50 could be chosen.

(c) Likely to be representative. Only unrepresentative if customers who come early in the day have different eating habits, as a group, from all customers or if, say, weekend customers had different eating habits from weekday customers.

3 **(a)** **(i)** A random, B systematic, C stratified.
(ii) A equally likely, B equally likely, C equally likely.

(b) No, not all subsets of size 128 of the population are possible.

(c) In a quota sample the names from each of the strata are selected by any convenient method. This is much easier and quicker, in practice, than selecting a random sample from each of the strata.

(d) The stratification is clearly relevant to the question in that the electors from the different strata are likely to have different views on local authority housing. Stratified sampling ensures that all strata are fairly represented in the sample and so is to be preferred to random sampling which does not.

(e) It is not obvious that the different strata will have different attitudes to the monarchy (although they may do). If there is no difference then extra work involved in stratified sampling is to no purpose.

4 **(a)** **(i)** Yes, each house has probability of $\frac{1}{4}$.
(ii) No, not all subsets of the population possible.

(b) Not all members of a household equally likely to answer the door, not all subsets of electors possible or chosen houses are not a random sample.

(c) Number residents 00 to 62; select two-digit random numbers; ignore repeats and >62; continue until seven numbers obtained; choose corresponding residents.

5 **(a)** C.

(b) A cluster, B stratified, D random.

(c) A no, B yes, C no (or depends on method of choosing sample), D yes.

(d) **(i)** B ensures branches fairly represented in sample and avoids possibility of bias.
(ii) C easier to carry out.

3 Approximating distributions

Note the following answers have usually been given to 3 sf and interpolation has been used in normal tables. As these are approximations 2 sf is sufficient and answers without interpolation are sufficiently accurate.

EXERCISE 3A

1 (a) 0.207, **(b)** 0.202, **(c)** 0.224, **(d)** 0.0077,
(e) 0.604, **(f)** 0.611, **(g)** 0.063.

2 0.047.

3 (a) 0.256,

(b) patients act independently.

4 0.699.

5 0.880.

EXERCISE 3B

1 (a) 0.944, **(b)** 0.813,
(c) 0.201, **(d)** 0.131,
(e) 0.345, **(f)** 0.652,
(g) 0.736, **(h)** 0.0995.

2 (a) 0.173, **(b)** 0.084.

3 (a) (i) 0.427, **(ii)** 0.199,
(b) (i) 0.258, **(ii)** 0.560, **(iii)** 0.182.

4 (a) (i) 0.341, **(ii)** 0.377
(b) (i) 0.011, **(ii)** 0.777, **(iii)** 0.845.

EXERCISE 3C

1 (a) 0.870, **(b)** 0.035, **(c)** 0.152, **(d)** 0.392,
(e) 0.123, **(f)** 0.089, **(g)** 0.168.

2 (a) 0.078, **(b)** 0.254.

3 (a) 0.111, **(b)** 0.181, **(c)** 0.076, **(d)** 0.091.

4 (a) (i) 0.205, **(ii)** 0.731
(b) (i) 0.941, **(ii)** 0.795.

EXERCISE 3D

1 (a) 0.208, **(b)** 0.505.

2 (a) (i) 0.850, **(ii)** 0.350
(b) 0.284, **(c)** 0.992.

3 (a) Poisson, **(b)** normal, **(c)** neither, **(d)** Poisson (use $p = 0.01$).
The binomial distributions would not be available in tables (although
they could be obtained from some calculators), the approximations
would enable tables to be used.

4 (a) 0.053, **(b) (i)** correct, **(ii)** not sensible.

5 (a) (i) 0.855, **(ii)** 0.976
(b) must be between 270 and 290, say 285.

6 (a) 0.173, **(b)** 11, **(c)** 0.176, **(d)** 44.

7 (a) 0.124 **(b)** 0.982.

4 Confidence intervals

To help you check your work, answers to this chapter have usually been given to 3 sf. However as all the calculations involve approximations, 2 sf would be sufficient. It would certainly be misleading to give your answers to more than 3 sf.

EXERCISE 4A

1 (a) $20.9 - 43.1$.
  (b) $17.4 - 46.6$.
  (c) $22.7 - 41.3$.
  (d) $2.1 - 4.3$.
  (e) $20.1 - 75.9$ (using AQA tables you need to derive the z value from Table 3 as Table 4 does not go as far as 0.9995).

2 (a) $18.4 - 39.6$.
  (b) Poisson approximated by normal; standard deviation only known approximately.
  (c) $40.3 - 75.7$.

3 (a) $13.6 - 32.4$. (b) $22.0 - 37.0$. (c) $21.9 - 29.3$. (d) $3.8 - 12.2$.

4 (a) $34.4 - 61.6$.
  (b) Confidence interval contains 40, so no convincing evidence that mean number of holidays sold per week has increased following the advertising campaign.
  (c) 40.7 to 60.3.
  (d) Lower limit of confidence interval exceeds 40. Taking the 2 weeks together provides evidence that the mean has increased following the advertising campaign.
  (e) Have assumed the Poisson distribution provides an adequate model for the number of holidays sold in the 2 weeks following the advertising campaign (in particular the mean is constant in these 2 weeks).

EXERCISE 4B

1 (a) $0.0281 - 0.0642$. (b) $0.0628 - 0.0999$.
  (c) $0.309 - 0.512$. (d) $0.128 - 0.272$.

2 (a) $0.079 - 0.174$.
  (b) The confidence interval contains 0.1 and so there is insufficient evidence to cast doubt on the claim.
  (c) The households Euan delivered to could be regarded as a random sample of all households.
  (d) Binomial approximated by normal, standard deviation only known approximately.

3 $0.171 - 0.276$.

4 (a) $0.278 - 0.478$.
  (b) Have assumed that the 90 eggs can be regarded as a random sample of all eggs laid on the farm.
  (c) The confidence interval contains 0.4 and so there no convincing evidence to doubt the farmer's claim.
  (d) When considering a normal approximation $p \leqslant 0.5$, and so in this case p would be the probability of an egg not being large. Estimate of np is 4 and this is too small for the normal to give a good approximation.

EXERCISE 4C

1 (a) $12.8 - 31.2$.

(b) $99 - 341$.

(c) (i) 71, (ii) 96.7.

(d) (i) $3.4\,\text{m}^2$, (ii) $1.4\,\text{m}^2$.

2 (a) $0.238 - 0.422$.

(b) $0.593 - 0.747$.

(c) Assume random sample of jars,

(d) (i) 48, (ii) 80.

(e) (i) 944, (ii) 488.

3 (a) 0.05, (b) 0.99, (c) 0.04, (d) 0, (e) 0.8, (f) 0, (g) 0.960.

5 Interpretation of statistics

EXERCISE 5A

1 (a) 105.9 million.

(b) (i) The number of sites seems to exhibit some random variation but no trend.

(ii) The number of screens shows an upward, approximately linear trend.
The average number of screens per site has increased steadily from just over two in 1988 to just over four in 1998.

(c) £2.77.

(d) The revenue per admission has increased from £1.89 to £3.64 from 1988 to 1998, an increase of 93%.
The retail price index has increased from 106.6 to 163.4, an increase of 53%. There has been a substantial increase in admission prices even allowing for the effect of inflation.

2 (a) 9.7 thousand tonnes.

(b) £7002 thousand.

(c) (i) Quantity of plaice landed has shown a downward trend with some indication that this trend has now levelled out.

(ii) Quantity of sand eels landed has shown a steep but erratic upward trend.

(d) (i) £743 per tonne, (ii) £542 per tonne.

(e) (i) £3260 per tonne, (ii) £4490 per tonne.

(f) Since the quantity of brill caught is given to only 1 sf the actual quantity is between 0.45 and 0.55. There could be an error due to rounding in the answers to (e) of up to about 10%.

(g) There was an increase in the RPI of 13% between 1992 and 1997. There has been a reduction of 27% in the average price of haddock. In real terms this reduction is even greater. There has been an increase of about 38% in the average price of brill. This comfortably exceeds the increase in the RPI (even allowing for possible rounding errors) and so represents a real increase in the price.

3 (a) The number of sites shows an upward trend. The rate of increase has been accelerating up to 1993. From 1993 to 1996 the rate of increase has reduced.

(b) The number of councils participating showed an upward, approximately linear trend until 1990 but then levelled off.

(c) Local government reorganisation reduced the number of councils in 1996.

(d) In 1979 only a small proportion of councils were participating whereas in 1985 a large proportion of councils were participating. It is therefore impossible to maintain the rate of increase in the number of councils participating.

(e) 1977 – 2.8, 1982 – 5.3, 1988 – 8.8, 1995 – 31.2.

As well as an increase in the number of councils participating there has also been a marked increase in the number of sites per participating council.

4 (a) 30.

(b) 22.

(c) The bigger the household income the bigger the proportion of women who have never smoked.

(d) As with women the bigger the household income the bigger the proportion of men who have never smoked. For any given household income group, the proportion of men who have never smoked is smaller than the proportion of women who have never smoked.

(e) 0.23 – 0.35, assume the sample may be treated as a random sample of men in this group.

(f) The figure for men from households with income £20 000 or more is given as 28% or 0.28. The interval calculated in **(e)** shows that the data provides no substantial evidence of a difference in the proportion of ex-cigarette smokers in these two income groups.

5 (a) 4.1 million tonnes

(b) (i) Downward, approximately linear trend.
(ii) Relatively little use of gas up to 1992 then a marked upward trend.
(iii) A relatively slow upward trend.

(c) 6.3 degrees. This sector would have been much larger in 1988 (32.2 degrees).

(d) Since the figure for hydro is given to only 1 sf the use in each of the years 1988 to 1991 may not have been the same but could have varied between 0.35 and 0.45 (million tonnes of oil equivalent). The increase from 0.4 to 0.5 might be almost entirely accounted for by rounding error.

(e) 1.9 million tonnes.

6 (a) (i) 4.5 g, **(ii)** 95.5 g, **(iii)** 91 g.
(b) (i) 0.0401, **(ii)** 0.00003.

(c) The average contents are less than the nominal quantity, more than 2.5% are non-standard. Requirements not met.

(d) More than 2.5% non-standard (8.5%), more than 0.1% inadequate (0.6%). Conditions not met.

1 (a) 29.3 – 54.7.

(b) 50 lies within the interval. There is no substantial evidence against the sales manager's claim.

2 (a) 30 671 000.

(b) (i) 165, **(ii)** 66 500.

(c) (i) The number of electors has shown an upward trend. Apart from the jump for the 1970 election due to the reduction of the minimum voting age, this has been a steady approximately linear increase.

(ii) The number of valid votes counted has also shown some upward trend. If the jump between 1970 and 1974 elections is ignored the trend largely disappears.

The upward trend of the number of valid votes counted has been less steep and more erratic than that of the number of electors.

3 (a) Yes, all have probability $\frac{1}{12}$ of being included in the sample.

(b) No, not all subsets of size 50 possible, for example two friends shopping together could not both be included.

(c) No, first 12 customers have probability $\frac{1}{12}$ of being included, the next 88 have no chance of being included.

(d) Likely to be representative. If customers entering later in the day had, as a group, different views than those entering early in the day the first sample would not be representative but the second would be. If customers coming on different days had, as a group different views then the samples would not be representative.

4

		1998				1999			
		Q1	Q2	Q3	Q4	Q1	Q2	Q3	Q4
Expenditure		163	265	409	186	165	293	469	196
Moving average			255.75	256.25	263.25	278.25	280.75		

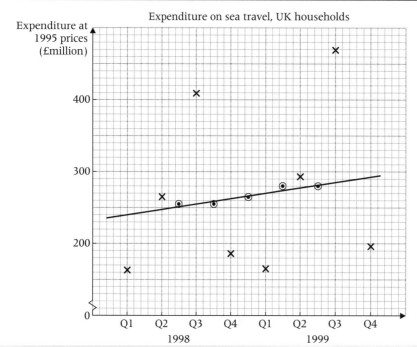

Expenditure on sea travel, UK households

Expenditure at 1995 prices (£million)

(b) Trend suggests moving average of about 300 for Quarter 1, 2000
Q1, 1998 is about 75 below trend, Q1 1999 is about 105 below
trend. $\frac{(75 + 105)}{2} = 90$. Forecast for Q1 2000 $300 - 90 = 210$

Forecast expenditure for Quarter 1, 2000 is £210 million (it would
be equally valid to base the estimate of seasonal effect on Q1 1999
only, which would lead to a forecast of about £195 million).

(c) The actual figure was well below the forecast. The forecast was
made on a very small amount of data and not much should be read
into the figures for one quarter but this does suggest that the
upward trend may not be continuing.

5 **(a)** 0.669.

(b) Probability may not be independent, e.g. if one member of a family
group does not turn up it may be more likely that other members
do not turn up.

(c) (i) Poisson, **(ii)** neither, **(iii)** Poisson.

Advancing Maths for AQA

The new route to A Level success

- **Advancing Maths for AQA** is the only series written exclusively for AQA.
- It's the only series written by the Senior Examining Team.

Advancing Maths for AQA guides you through the course in a clear and logical way, covering only the topics you need to study. The books are filled with clear explanations, key points and graded examples, which build on the basics to bring you up to exam level. And it's easy to check your progress too: with 'test yourself' sections and a full exam paper you can really work on your problem areas. Plus, with tips from the examiners on how to achieve more, you can get the marks that you deserve.

Why would you need anything else?

> **To see any of the following titles FREE for 60 days or to order your books straight away call Customer Services on 01865 888068**

Pure Mathematics 1 (P1) 0435 513001	Mechanics 3 (M3) 0435 513087	Statistics 5 (S5) 0435 513168
Pure Mathematics 2 (P2) 0435 513044	Mechanics 4 (M4) 0435 513095	Statistics 7 (S7) 0435 513222
Pure Mathematics 3 (P3) 0435 513028	Mechanics 5 (M5) 0435 513109	Statistics 8 (S8) 0435 513230
Pure Mathematics 4 & 5 (P4 &P5) 0435 513036	Mechanics 6 (M6) 0435 513117	Discrete Mathematics 1 (D1) 0435 513184
Pure Mathematics 6 (P6) 0435 513052	Statistics 1 (S1) 0435 513125	Discrete Mathematics 2 (D2) 0435 513192
Pure Mathematics 7 (P7) 0435 51301X	Statistics 2 (S2) 0435 513133	
Mechanics 1 (M1) 0435 513060	Statistics 3 & 6 (S3 & S6) 0435 513141	
Mechanics 2 (M2) 0435 513079	Statistics 4 (S4) 0435 51315X	